CHLOROPHYLL

Edited by **Eduardo Jacob-Lopes, Leila Queiroz Zepka** and **Maria Isabel Queiroz**

Chlorophyll
http://dx.doi.org/10.5772/65594
Edited by Eduardo Jacob-Lopes, Leila Queiroz Zepka and Maria Isabel Queiroz

Contributors

Jericó Jabín Bello Bello, Eduardo Martínez Estrada, Carlos Alberto Cruz Cruz, Juan Antonio Pérez-Sato, Ana Carolina Mazzuco, Paula Kasten, Lucia Guidi, Massimiliano Tattini, Marco Landi, Abdul Kariem Arof, L.P. Teo, Andreas Petsas, Maria Vagi, Eduardo Jacob-Lopes, Andrêssa Fernandes, Mariany Deprá, Leila Queiroz Zepka, Maria Isabel Queiroz, Felix Nchu, Kerwin Lefever, Charles P Laubscher, Patrick A. Ndakedemi

Published by InTech
Janeza Trdine 9, 51000 Rijeka, Croatia

Publishing Process Manager Nina Kalinic
Technical Editor SPi Global
Cover InTech Design team

First published May, 2017
Printed in Croatia
Legal deposit, Croatia: National and University Library in Zagreb

Additional hard copies can be obtained from orders@intechopen.com

Chlorophyll, Edited by Eduardo Jacob-Lopes, Leila Queiroz Zepka and Maria Isabel Queiroz
p. cm.
Print ISBN 978-953-51-3107-6
Online ISBN 978-953-51-3108-3

For EU product safety concerns:
IN TECH d.o.o., Prolaz Marije Krucifikse Kozulić 3, 51000 Rijeka, Croatia,
info@intechopen.com or visit our website at intechopen.com.

Meet the editor

Editor, Prof. Eduardo Jacob-Lopes is currently an associate professor at the Department of Food Technology and Science, Federal University of Santa Maria. He graduated with a master's degree in Food Engineering in Federal University of Rio Grande do Sul, doctorate degree in Chemical Engineering from the State University of Campinas, and postdoctoral at the State University of Campinas. He has more than 15 years of teaching and research experience. He is a technical and scientific consultant of several companies, agencies, and scientific journals. He has 360 publications/communications which include 3 books, 17 book chapters, 60 original research papers, and 280 research communications in international and national conferences and has registered 7 patents. His research interest includes environmental biotechnology with emphasis on microalgal biotechnology.

Co-Editor, Prof. Leila Queiroz Zepka is currently an associate professor at the Department of Food Technology and Science, Federal University of Santa Maria. She graduated with a master's degree in Food Engineering in Federal University of Rio Grande do Sul and a doctorate degree in Food Science from the State University of Campinas. She has more than 15 years of teaching and research experience. She is a technical and scientific consultant of several companies, agencies, and scientific journals. She has 360 publications/communications which include 3 books, 14 book chapters, 50 original research papers, and 200 research communications in international and national conferences, and has registered 4 patents. Her research interest includes microalgal biotechnology with emphasis on biomolecules.

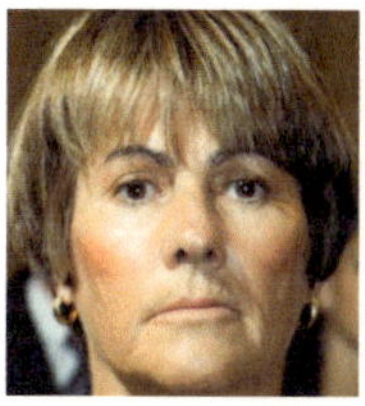

Co-Editor, Prof. Maria Isabel Queiroz is currently a senior professor at the Department of Chemistry and Food, Federal University of Rio Grande. She graduated in oceanography at the Federal University of Rio Grande, with a master's degree in Food Technology in the State University of Campinas and a doctorate degree in Biotechnology from the Federal University of Pelotas. She has more than 30 years of teaching and research experience. She is a technical and scientific consultant in several companies, agencies and scientific journals. She has more than 500 publications/communications which include 7 books, 7 book chapters, 100 original research papers and 350 research communications in international and national conferences. Her research interest includes microalgal biotechnology with emphasis on bioprocesses.

Contents

Preface

Chlorophylls are fundamental molecules of life as we know in our planet. Chlorophylls are probably the most important of all natural pigments, occurring in plants and microorganisms. Chlorophylls are responsible for many spectacular shows of colour in nature, and, as a consequence, their distribution has been thoroughly examined. Chlorophylls are also of commercial importance as they have been used for a number of years as colourants and antioxidants and in the monitoring of agricultural production and ocean primary productivity. In recent years, great advances have been made in understanding the biological functions of chlorophylls. This is especially true for photosynthetic organisms, where the chlorophylls serve one important function, namely, light-harvesting. Chlorophylls are key components of the photosynthetic apparatus and have a vital role in the evolution of photosynthetic organisms. Chlorophylls are also essential for the structural organisation of the photosynthetic apparatus. Recent advances in chlorophyll photochemistry have been impressive, and clear roles for chlorophylls in photosynthesis are now evident. Thus, the main aim of this book is to bring together the key aspects of the biology, biochemistry and chemistry of chlorophylls in photosynthetic organisms.

Prof. Eduardo Jacob-Lopes
Federal University of Santa Maria
Santa Maria, Brazil

Prof. Leila Queiroz Zepka
Federal University of Santa Maria
Santa Maria, Brazil

Prof. Maria Isabel Queiroz
Federal University of Rio Grande,
Rio Grande, Brazil

Chapter 1

Introductory Chapter: Chlorophyll Molecules and Their Technological Relevance

Maria Isabel Queiroz, Andrêssa Silva Fernandes, Mariany Costa Deprá, Eduardo Jacob-Lopes and Leila Queiroz Zepka

Additional information is available at the end of the chapter

http://dx.doi.org/10.5772/67953

Chlorophylls were the major tool used, long before science was able to recognize the existence of plants and animals; our predecessors identified this distinction and slowly provided technological evidence that chlorophyll was responsible for the green color of the plants.

The name chlorophyll was proposed by Pierre Joseph Pelletier and Joseph Bienaimé Caventou in 1818 to designate the Green substance that could be extracted from the leaves with the aid of alcohol [1–3]. However, only about a century later, in 1911, the study of this interesting molecule gained more importance by Richard Willstätter, who obtained for the first-time pure chlorophyll and established its correct molecular formula $C_{55}H_{70}MgN_4O_6$ [2]. A year later, Willstätter showed that chlorophyll obtained from a wide variety of sources was a mixture of two compounds, chlorophyll-a and chlorophyll-b, to which he assigned the correct molecular formula $C_{55}H_{70}MgN_4O_5$ and $C_{55}H_{70}MgN_4O_6$, respectively. Nevertheless, the detailed understanding of the structures of these molecules was the result of the studies of Fischer and collaborators, who were the first to delineate the structure of porphyrin ring [2–5].

These compounds are complex organic molecules formed by derivatives of porphyrin, a macrocyclic structure, asymmetric, totally unsaturated. Basically, chlorophyll molecules are conjugated tetrapyrroles to which a cyclopentanone ring jointly with the third ring linked together by bridges methylene, with the central atom, magnesium (**Figure 1**). This structure also contains at the C-17, a propionic acid chain esterified with the phytol, diterpene alcohol [3, 6–8].

Chlorophyll-b differs from chlorophyll-a by the presence of aldehyde residue instead of the methyl group at position 7. The synthesis of this compound is given by the oxidation of the methyl group to aldehyde [9, 10] through the enzyme oxygenase, which catalyzes this conversion [11].

Structurally, the chlorophyll molecules differ from each other, due to the degree of saturation of the pyrrolic rings. For example, chlorophyll-c contains fully unsaturated phytoporphyrin (double alloys C17–C18), whereas the other chlorophylls are C17–C18, phytochlorins (**Figure 1**). These differences in chlorophyll macrocycle saturation have profound consequences on the absorbance spectrum. For example, chlorophyll-a, chlorophyll-d, and chlorophyll-f phytochlorins have approximately equal intensities of absorption in blue, red, and green. On the other hand, the phytoporphyrins of chlorophyll-c absorb themselves weakly in red and more intensely around 450 nm [6, 12, 13].

The spectral differences of the chlorophyll molecules are reflected in the key in which bodies are present that the chlorophyll-a appears more bluish green, chlorophyll-b bright green, chlorophyll-c yellowish green, chlorophyll d bright forest green, and chlorophyll-f emerald green [6]. This diversity of shades of green makes these organisms potential source of natural colorants [14].

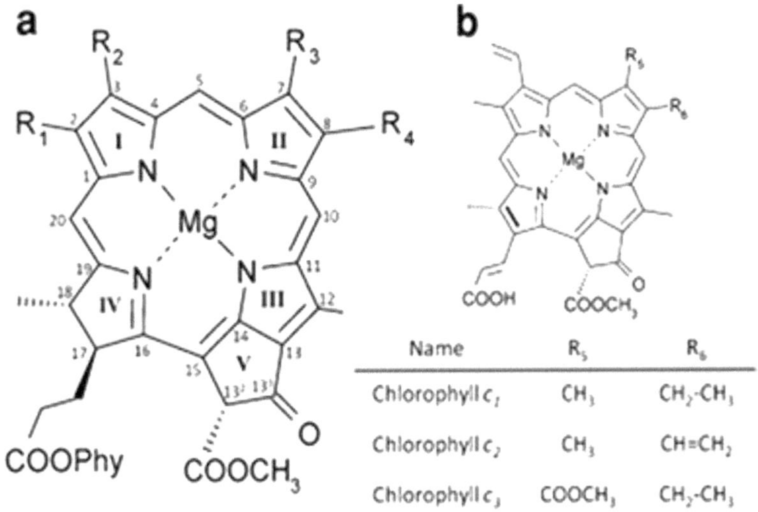

Name	R_5	R_6
Chlorophyll *c₁*	CH_3	CH_2-CH_3
Chlorophyll *c₂*	CH_3	$CH=CH_2$
Chlorophyll *c₃*	$COOCH_3$	CH_2-CH_3

Name	R_1	R_2	R_3	R_4
Chlorophyll *a*	CH_3	$CH=CH_2$	CH_3	CH_2-CH_3
Chlorophyll *b*	CH_3	$CH=CH_2$	CHO	CH_2-CH_3
Chlorophyll *d*	CH_3	CHO	CH_3	CH_2-CH_3
Chlorophyll *f*	CHO	$CH=CH_2$	CH_3	CH_2-CH_3
8-vinyl Chlorophyll *a*	CH_3	$CH=CH_2$	CH_3	$CH=CH_2$
8-vinyl Chlorophyll *b*	CH_3	$CH=CH_2$	CHO	$CH=CH_2$

Figure 1. Structure of the side chain variables of chlorophyll molecules.

In addition, a chlorophyll molecule consists of two distinct parts: hydrophile, the macrocycle, and a hydrophobic part, phytol chain. The most hydrophilic segment of the macrocycle is the cyclopentanone ring and the propionic ester group (17 positions). The hydrophilic and hydrophobic character directly influences the choice of solvent for extraction, which, from a quantitative or qualitative point of view, is an extremely important aspect when it focuses on the technological relevance of chlorophyll molecules [15–17].

Given this scenario, analysis techniques such as liquid chromatography coupled to ultraviolet and mass spectrum allowed demonstrating that currently a total of five molecules, so-called chlorophyll, have been found in nature. All of them occur in the cyanobacterium, which is cosmopolitan microalgae, prokaryotic, Gram negative, with metabolism preferentially photoautotrophs, able to perform photosynthesis oxygen production, eukaryotic algae, and higher plants. These microorganisms demonstrate considerable biocatalytic potential in biotechnological processes due to their robustness and require nutritional simplicity [18]. Chlorophyll-a is present in all organisms that perform photosynthesis and chlorophyll-b oxygen (**Figure 1**), which are the most abundant pigment in green algae, is found in higher plants in nature, and other similar structures are used as accessory pigments in photosynthetic process, called chlorophyll-c, chlorophyll-d, chlorophyll-e, and chlorophyll-f [6–8, 13]. These specific pigments present in the cells of the photosynthetic organisms influence efficiency of dispersion of light, because each species has its own distinct characteristic pigment [9].

Chlorophyll-a is the most abundant and important structure of all chlorophylls, corresponding to approximately 75% of the green pigment found in nature. This molecule has bioactive properties with extensive use in pharmaceutical and food industry, is considered a high added value, and may reach high values on the market. The bioactive potential of the cells is associated with a complex structure of conjugated connections in pyrrolic rings, allowing an oxidant and antioxidant, acting without free radical sequestration [7]. In addition, it is related to nutraceutical properties, playing an important role in health, through its anti-inflammatory and vascular constrictor, which makes this molecule an important ingredient in food formulations, cosmetics, and drugs [16].

Another molecule that differs from chlorophyll-a, to present the first pyrrole ring, a substitution of the vinyl C3 group for the chlorophyll-a molecule for a formyl group, was called chlorophyll-d. This molecule was discovered 70 years ago, but only in 1996, an important property of this molecule was observed [6, 9, 13], performing a function equivalent to chlorophyll-a in oxygenic photosynthesis [7, 13]. The properties of chlorophyll-d, like its photosynthetic power associated with the difference in wavelength of absorption in relation to chlorophyll-a, make this a potential component molecule to be employed in a biorefinery. In the last years, it has been argued that there is a strong interest in the development of technologies for the production of natural molecules [16, 19, 20].

On the other hand, chlorophyll biomolecules are highly prone to change, because their chemical structure is rich in double bonds combined. In this sense, in order to increase stability, chlorophyll undergoes changes in its molecule, replacing the atom of Mg^{2+} for Cu^{2+}, resulting in the call copper chlorophyllin, which is stable and can be used in hydro- or lipo-soluble formulations [21].

In summary, the broad application of chlorophyll makes it a molecule of great importance for the global market. However, there is still doubt about its application due to its chemical instability. In this way, the extension of studies to elucidate the open gaps will be of extreme relevance to expand the scientific knowledge base and its relation to the industrial application.

Author details

Maria Isabel Queiroz[1], Andrêssa Silva Fernandes[2], Mariany Costa Deprá[2], Eduardo Jacob-Lopes[2] and Leila Queiroz Zepka[2*]

*Address all correspondence to: zepkaleila@yahoo.com.br

1 School of Chemistry and Food, Federal University of Rio Grande, UFRGS, Rio Grande, RS, Brazil

2 Food Science and Technology Department, Federal University of Santa Maria (UFSM), Santa Maria, RS, Brazil

References

[1] Pelletier, F.; Caventou, J. B. Sur la matièri verte dês feuilles. Annales de C himieet de Physique, 9, 194–196, 1818.

[2] Gossauer, A.; Engel N. Chlorophyll catabolisms—Structures, mechanisms conversions. Journal of Photochemistry and Photobiology, 32, 141–151, 1996.

[3] Schoefs, B. Chlorophyll and carotenoids analysis in food products properties of the pigments and methods of analysis. Trends in food Science & Technology, 13, 361–371, 2002.

[4] Fischer, H.; Orth. H. Die chemie des pyrrols. Akaemie Verlagsgesellschaft, Vol. 2, Part 1, Leipzig, Germany, 1937.

[5] Fischer, H.; Stern, A. Die chemies des pyrrols. Akademie Verlagsgesellschaft, Part 2, Leipzig, Germany, 1940.

[6] Chen, M.; Blankenship, R. E. Expanding the solar spectrum used by photosynthesis. Trends in Plant Science, 16, 488–496, 2010.

[7] Chen, M.; Yaqiong, L.; Birch, D.; Willows, R. D. A cyanobacterium that contains chlorophyll-f a red absorbing photopigment. FFBS Letters, 586, 3249–3254, 2012.

[8] Fu, W.; Magnusdótter, M; Palson, B. UPLC – UV – MS analysis for quantification and chlorophyll species in algae. Analytical and Bioanalytical Chemistry, 404, 3145–3154, 2012.

[9] Harada, J.; Mizoguchi, T.; Tsukatani, Y.; Noguchi, M.; Tamiaki, H. A seventh Bacterial chlorophyll driving a large light-harvesting antenna. Scientific Reports, 2(671), 1–5, 2012.

[10] Yen, H. W.; Hu, I. C.; Chen, C. Y.; Ho, S. H.; Lee, D. J.; Chang, J. Microalgae based biorefinaria – from biofuels to natural products. Bioresource Technology, 35, 166–174, 2013.

[11] Xu, W.; Tang, H.; Wang, Y.; Chitins, P. Proteins of the cyanobacterial photosystem I. Biochimica et Biophysica Acta, 1507, 32–40, 2001.

[12] Zapata, M.; Rodriguez, A.; Garrido, J. L. Separation of chlorophyll and carotenoids from marine phytoplankton a new HPLC method using a reversed phase C8 column and pyridine containing mobile phases. Marine Ecology Progress Series, 195, 29–45, 2000.

[13] Loughlin, P.; Lin, Y.; Chen, M. Chlorophyll d and Acaryochloris marina: current status. Photosynthesis Research, 116(2–3), 277–293, 2013.

[14] Pelayo, R. D.; Galhardo-Guerreiro, L; Homero, M. Chlorophyll and carotenoid pigments in the peel and flesh of commercial Apple fruit varieties. Food Research International, 65, 272–281, 2014.

[15] Rosseló-Soto, E.; Galanakis, C. M.; Brncic, C. M.; Orliens, V. Trugillo, F. J. R.; Mawson, R.; Barba, F. J. Clean recovery of antioxidant compounds from plants food, byproducts and algae assisted by ultrasound processing. Modelling approaches to optimize processing conditions. Trends in Foods Science & Technology, 42, 134–149, 2015.

[16] Fernandes, A.; Nogara, G.; Menezes, C. R.; Cichoski, A. J.; Mercadante, A. Z.; Jacob-Lopes, E.; Zepka, L. Q. Identification of chlorophyll molecules with peroxyl radical cavenger capacity in microalgae *Phormidium autumnale* using ultrasound-assisted extraction. Food Research International, *In Press*, 2016.

[17] Streit, N.; Merida, L. G. R.; Zepka, L. Q.; Jacob-Lopes, E.; Queiroz, M. I. Production de pigmentos por Aphanothece microscopica näglia partir de resíduos industrialis lácteos. Ingeniare Revista Chilena de Ingenieria, 25, 11–18, 2017.

[18] Guiry, M. D.; Guyri, G. M. Algae Base: world–wide electronic publication National University of Ireland. Available at: http:\\www.algaebase.org. Accessed Oct 13, 2016.

[19] Queiroz, M. I.; Hornes, M. O.; Silva-Manetti, A. G.; Zepka, L. Q.; Jacob-Lopes, E. Fish processing wasterwater a platform of the microalgal biorefineries. Biosystem Engineering, 115, 95–202, 2013.

[20] Kim, S.; Kim, M. S.; Lee, B. Y.; Lee, P. Generation of structurally novel short carotenoids and study of their biological activity. Scientific Reports, 6, 1–11, 2016.

[21] Sant'Anna, V., Gurak, P. D., Marczak, L. D. F., Tessaro, I. C. Tracking bioactive compounds with colour changes in foods–A review. Dyes and Pigments, 98(3), 601–608, 2013.

Chapter 2

Chlorophyll-a and the Supply Side Ecology: Lessons from the Rocky Shores

Ana Carolina de Azevedo Mazzuco and
Paula Kasten

Additional information is available at the end of the chapter

http://dx.doi.org/10.5772/68044

Abstract

The aims of this study were to summarize and describe the influences of phytoplankton on the larval cycle of rocky shore invertebrates, and to assess the relationship between fluctuations in chlorophyll-a concentration and the rates of larval processes. We carried out a mini review of the published data regarding the theme of the chapter, in which we described the ecological trends for the most common taxa and key species at small and larger spatiotemporal scales. The following topics were addressed: (i) the influence of phytoplankton on larval development, rhythms of larval release, larval quality, larval transport, settlement, and recruitment; (ii) the relationships between variations in chlorophyll-a concentration and the rates of larval processes; (iii) climate change on phytoplankton larva dynamics. The information presented here highlights the role of phytoplankton on rocky shore communities, as well as the importance of chlorophyll-a as a tool for modeling and forecasting the supply side ecology in rocky shore communities.

Keywords: phytoplankton, chlorophyll-a, supply side ecology, marine invertebrates, rocky shores, benthic-pelagic coupling

1. Introduction

Larval supply is the main source of new individuals to the populations of rocky shore invertebrates [1–3]. In these communities, larval success regulates how energy is transferred through the trophic web [4–6]; consequently, variations in the supply of propagules are the basis of trophic interactions at rocky shores [7, 8]. Since phytoplankton is the main food source for planktonic larvae of marine invertebrates [9], variations in phytoplankton biomass and diversity have significant influences on the larval cycle. Larval responses to the variability in phytoplankton abundance and diversity are species-specific. Larval fitness is influenced by environmental conditions experienced by adults and larvae [10, 11]. The effects of phytoplankton on larval dynamics depend on the phase of larval development [12–15] and may be stronger when variations in phytoplankton occur on temporal scales that larvae or breeding adults are able to respond [16]. The direct interaction between phytoplankton and the larval stages have short-term consequences for larval dynamics (e.g., Ref. [14]), and it might have long-term effects as well. Because of that, variations in the rates of the ecological processes of rocky shore invertebrates are commonly correlated with fluctuations in chlorophyll-a concentration in the ocean (e.g., Refs. [17–20]). These numerical relationships are important tools to ecological modeling, and may be used to improve stock management in some extent [21].

2. The role of phytoplankton blooms in reproduction timing and in the rhythms of larval release

In the rocky shore communities, filter feeders depend greatly on phytoplankton as their main source of food and its consumption results in energy for growth and reproduction [22]. It is common to find larger animals with higher fecundity rates at rocky shores located in areas of high primary productivity, as a response to the higher concentrations of phytoplankton, and thus, food availability [19, 23–25]. Different types of phytoplankton present distinct physiological qualities as food particles [26], thereby both the amount of phytoplankton in the water column and their diversity influence the reproductive traits in marine invertebrates.

But not only adults on the rocky shore depend on phytoplankton in order to survive, larvae produced by those organisms also rely on these microorganisms to develop and reach the juvenile phase [27]. As evolution drives maximum reproductive activity to happen when environmental conditions are the best for offspring development, food availability is one of the most important factors regulating reproduction and allowing adults to produce viable offspring. Thus, it is common to observe peaks of larval release by rocky shore invertebrates synchronized with phytoplankton blooms (e.g., Refs. [28, 29]). Some metabolites produced by phytoplankton are signs of favorable environmental conditions for the larval development, trigging the spawning activity of green sea urchins and blue mussels, for instance Ref. [28]. These animals perceive such chemical compounds as an indication of good food abundance, so synchronizing the timing of larval release with high abundance of phytoplankton would promote higher offspring survival. Barnacles, on the other hand, just need a physical contact with phytoplankton cells to trigger their spawning activity, and larger the phytoplankton cell is, the stronger is the response [28].

Therefore, the presence of phytoplankton may overcome other environmental factors in the regulation of reproduction timing and larval release [30]. Spring and summer are the main reproductive periods for rocky shore invertebrates at temperate and upwelling regions [31], as it is during these seasons that phytoplankton blooms occur. Mussels from the Baltic sea, for example, start to develop their gonads when temperature starts to drop in the beginning of winter; but its maturation and ripening processes proceed in a way that the animals are ready to reproduce at the same time that phytoplankton blooms occur in the beginning of spring [32]. Some barnacles are even able to maintain their fully developed nauplii in the mantle cavity until a high abundance of phytoplankton is perceived by the adults and only then, the nauplii will be released, a strategy that enhances the offspring survival due to the higher chance of facing a favorable feeding environment [33].

Similar reproductive timing was registered in the Indian coasts, where phytoplankton blooms occur during the monsoons and barnacles spawn their nauplii short after a break of the monsoon conditions [34]. However, these are not the best conditions for nauplii development, as these breaks stop and unfavorable monsoon conditions for larval development return soon after. Such misleading cue could result in lower recruitment rates for barnacles in this region. In subtropical coasts, peaks of larval production in intertidal barnacles are also preceded by high concentrations of chlorophyll-a in the water column [35]. On the daily scale, phytoplankton diversity might be as important as biomass in the regulation of larval release [36]. The presence of phytoplankton may overcome other environmental factors known to act as synchronization cues for reproduction timing and larval release [30].

3. How do changes in phytoplankton affect larval development from release to competency?

As seen in the previous section, phytoplankton has an important role in the reproductive success of marine invertebrates inhabiting the rocky shores. Part of this reproductive success involves the survival of larvae up to the juvenile stage, and a successful return to the benthic habitat is essential to the maintenance of rocky shore populations [2, 37]. It is straightforward to think that larval development is strictly linked to changes in phytoplankton community, since these cells are the main food items for marine planktotrophic larvae [9]. Because of that, the physiological quality of a larva would be determined in the plankton during its development and influenced directly by the phytoplankton in the water column. However, phytoplankton may change larval physiological quality much before that same larva is produced, through maternal effects, that is, when maternal individuals have the capacity to perceive the environment and manipulate the energy allocated for propagule production [38].

The amount of energetic reserves allocated to each propagule produced depends on the amount of energy the maternal individual can provide to its offspring. This capacity, in turn, is limited by the food available for the mothers, their perception of it, and their competency to gather and assimilate energy [38, 39]. For those marine organisms that produce lecithotrophic larvae, maternal effects are extremely important for shaping larval physiological quality because these larvae depend exclusively on the energetic resources from embryogenesis to survive [40]. If food ration is low, mothers can either preserve the energy acquired for their own metabolism and produce lower quality larvae (a selfish strategy, Ref. [39]) or invest all energy possible into their propagules, enhancing the survival potential of that higher quality larvae (an anticipatory strategy, Ref. [39]). In a scenario where maternal individuals are feeding mainly on phytoplankton, as the majority of filter feeding invertebrates in the rocky shores are, it is possible to understand the effect that oscillations in the quantity and type of phytoplankton available for these animals to feed has on larval quality.

However, most invertebrates that inhabit the rocky shores produce planktotrophic larvae. These larvae are submitted to transport and dispersion; they will feed in the plankton and will probably not experience the same conditions of the maternal environment, hypothetically reducing the necessity of energy transfer from mother to larvae. Thus, one could assume that the food environment experienced by mother would not impact the quality of the larvae produced. Interestingly, few authors have shown that, under stressful temperatures and low phytoplankton concentrations, maternal individuals of a tropical barnacle are able to manipulate the transfer of different types of fatty acids to their nauplii, a possible strategy to guarantee higher survival rates until this same nauplii encounters better food conditions in the water column [41]. Variations in the amount and type of phytoplankton available for planktotrophic larvae during development cycle interfere in the different larval traits, including in the success of metamorphosis into the juvenile stage. Larvae of gastropods [15, 16, 42], bivalves [36, 43, 44], and barnacles [45, 46] vary in size, development rate, and survival to the juvenile stage, in direct association with the quality and amount of phytoplankton offered them during their development.

Larvae must be able to survive from pelagic to benthic conditions and return to the rocky shore communities, in order to reach the adult phase. Settlement success and post-settlement survivorship are also matters of larval history [12, 15, 21], and many more. Contrary of what has been accepted for a long time, settlement of larvae in the benthic environment, and its metamorphose to the juvenile stage do not result in a "new beginning" for those individuals, but the feeding conditions experienced by larvae and its results on their physiological quality can be carried over to the next stage, and those individuals who faced low phytoplankton concentrations during its life in the plankton might become juveniles with lower growth and survival potential, influencing directly on the fate of that population [46–51].

4. Larval transport, settlement, and recruitment

Phytoplankton and larval abundances are sometimes controlled by the same oceanographic processes. Phytoplankton grows and reproduces under very specific environmental conditions, driven mainly by turbulence and nutrient availability [52]. Ocean movements, such as turbulence, vertical mixing, and currents, also affect larval abundance at small (e.g., Ref. [53]) and larger scales (e.g., Ref. [54]). Marine larvae take advantage of meso- and large-scale oceanographic features for transport and dispersion. These larvae have different responses depending on the velocity at that depth, assuming a specific swimming or orientation pattern (e.g., Ref. [55]). Besides, larvae are able to control their position in the water column and move together with the main current at that specific depth [56–58], what in turn might result in variability of larval supply in time and space [59]. Some oceanographic features that accumulate and transport marine invertebrate larvae are responsible for disturbing phytoplankton as well. For example, upwelling currents, which cause phytoplankton blooms by injecting cold nutrient-rich waters in the photic zone, may move larvae of rocky shore invertebrate to shallower waters (e.g., Refs. [60, 61]). Storms are other meteorological-oceanographic phenomena that disturb both chlorophyll-a concentration at the nearshore environments (e.g., Ref. [62]) and the larval abundances close to the rocky shores [63].

Settlement is a function of larval supply [64]. Consequently, successful settlement relies on larvae, which need to find suitable settlement sites and be able to metamorphose. In this phase of the larval cycle, biochemical and physical cues either stimulate or block settlement. The presence of biofilm on the rocks is very important for settling larvae, in particular for the sessile larvae, because biofilm may define if that is a favorable settlement spot. Biofilm characteristics control larval behavior during settlement [65]; as a result, settlement rates and the chlorophyll-a content in the biofilm are correlated [66]. Settlement may also be correlated with fluctuations in chlorophyll-a concentration just as a consequence of the coupling between phytoplankton blooms and larval release [12, 28]. When the latter situation is true, fluctuations in chlorophyll-a concentration and variations in settlement rates are time lagged in several days [35], what may depend on the time that the larva takes to fully develop. On the other hand, if larval supply and phytoplankton dynamics are controlled by the same features, as it was explained in the previous paragraph, peaks in chlorophyll-a concentrations and settlement rates will occur simultaneously (e.g., Ref. [20]).

Recruitment rates are regulated by fluctuations in the pelagic environment affecting larval supply [67]. Recruitment success means that settled larvae survived until they are able to reproduce. In the post-settlement period, phytoplankton availability in the benthos and pelagial can control the survivorship of settlers in rocky shore communities. Although most early recruits of rocky shore invertebrates are filter feeders, they do not have the same diet and they may be very selective [68], choosing determinate phytoplankton species as food items depending on their size. Changes in the phytoplankton community might benefit one or the other species depending on their feeding behavior [68]. Although the relationship between recruitment and chlorophyll-a concentration is influenced by species-specific characteristics, information on this subject is still relatively scarce for rocky shore invertebrates. Small- and large-scale spatial variability in recruitment of rocky shore invertebrates are related to local and regional gradients of chlorophyll-a concentration in the surface waters. Geographic barriers that restrict phytoplankton abundance are also responsible for setting geographical limits for recruitment at the rocky shores. Recruitment rates may vary in several orders of magnitude among regions and sites, potentially due to persistent gradients in phytoplankton availability, and in turn gradients in chlorophyll-a concentration (e.g., Refs. [69, 70]. Even sites within the same bay or just less than 1 km apart may present high contrasts in recruitment rates as a consequence of differences in the phytoplankton dynamics [71].

5. The numerical relationships between chlorophyll-a concentration and larval processes

Phytoplankton is a limiting resource to the survival of marine invertebrate larvae, as it was described throughout the chapter; consequently, chlorophyll-a concentration is a key factor regulating larval dynamics in rocky shore communities. Variations in larval processes and fluctuations in chlorophyll-a concentration tend to be highly correlated (e.g., trends of recruitment rates [69]). These correlations could be incorporated to ecological and numerical models to predict larval processes based on the values of chlorophyll-a concentration in the water (e.g., Ref. [72]). Although there are daily measurements of chlorophyll-a concentration in the ocean surfaces at a global scale, the levels of correlation between chlorophyll-a and larval dynamics are described only for a few species and some coastal areas.

Trends may be divided in groups according to the relationship between larval and phytoplankton dynamics. If the oceanographic processes promoting larval supply and settlement are also responsible for enabling phytoplankton growth and reproduction, variations in larval processes and in chlorophyll-a concentration may be positively correlated. On the other hand, if larval supply and settlement are enabled by less favorable conditions for phytoplankton, the fluctuations in the rates of larval processes may be negatively related to the concentrations of chlorophyll-a. Evidences of both trends were registered for rocky shore invertebrates in several regions [20, 21, 73]. Although the oceanographic and ecological processes that affect community dynamics are similar at the rocky shores, the correlation degrees between phytoplankton abundance and larval processes vary among sites and taxa. Correlations are stronger when reproduction and larval processes are regulated by the same mechanisms controlling phytoplankton blooms. For instance, in upwelling regions, these correlations are expected to be stronger [74], but may not be significant depending on the site (e.g., Ref. [75]). Barnacle and mussel recruits that occupy the same intertidal zone are not necessarily affected by fluctuations in chlorophyll-a concentration in similar ways, even presenting opposite trends in recruitment [21].

6. Climate change on phytoplankton larval dynamics

Climate change has important consequences for benthic-pelagic dynamics. Global warming has already caused alterations in the patterns of sea surface temperature and ocean currents, which in turn directly influenced the trends of phytoplankton abundance. Larvae and recruits of rocky shore invertebrates have to cope with such alterations in food availability concomitant to other climatic changes. The effects of phytoplankton and other climatic factors, such as water temperature, tend to be synergic [76]. Global warming conditions might not be positive for marine invertebrate larvae which, on one hand, survive under a wide range of conditions, but their fitness is highly influenced by changes in food availability. Short- and long-term consequences of climate change on phytoplankton larval dynamics were already detected for rocky shore communities. On the scale of decades, longer events of upwelling in the recent 20 years doubled the recruitment rates in some shores [77]. Results showed that, in small scale conditions, variability in phytoplankton has different effects on larval performance under different levels of climate change (Kasten, personal communication). However, how species will respond to multiple factors under *in situ* oceanic climatic conditions are hard to forecast, since information in larval dynamics are not available for most species and rocky shore systems.

7. Final considerations

Phytoplankton has a high regulatory potential in larval dynamics in the rocky shore communities. Rates of larval processes in rocky shore invertebrates are highly correlated with spatiotemporal fluctuations in chlorophyll-a concentration in the sea surfaces. The role of phytoplankton in larval dynamics at the community levels is not known, because information for most species is incipient. It is important to highlight that scientific improvements are needed to allow that use of variations chlorophyll-a concentration as a tool for modeling and forecasting the supply side ecology in rocky shore communities.

Author details

Ana Carolina de Azevedo Mazzuco[1,2]* and Paula Kasten[2]

*Address all correspondence to: ac.mazzuco@me.com

1 Federal University of Espírito Santo, Dept. of Oceanography, Vitória, ES, Brazil

2 Federal University of São Paulo, Institute of Ocean Sciences, Santos, SP, Brazil

References

[1] Roughgarden J, Iwasa Y, Blaxter C (1985) Demographic theory for an open population with space-limited recruitment. Ecology 66:54–67. doi: 10.2307/1941306

[2] Roughgarden J, Gaines S, Possingham H (1988) Recruitment dynamics in complex life cycles. Science 241:1460–1466. doi: 10.1126/science.11538249

[3] Gaines S, Bertness M (1992) Dispersal of juveniles and variable recruitment in sessile marine species. Nature 360:579–580. doi:10.1038/360579a0

[4] Bustamante RH, Branch GM, Eekhout S, Robertson B, Zou-tendyk P, Schleyer M, Dye A, Hanekom N, Keats D, Jurd M, McQuaid C (1995) Gradients of intertidal productivity around the coast of South Africa and their relationships with consumer biomass. Oecologia 102:189–201. doi: 10.1007/BF00333251

[5] Bustamante RH, Branch GM, and Eekhout S (1995) Maintenance of an exceptional intertidal grazer biomass in South Africa: subsidy by subtidal kelps. Ecology 76:2314–2329. doi: 10.2307/1941704

[6] Nielsen KG, Navarrete SA (2004) Mesoscale regulation comes from the bottom-up: intertidal interactions between consumers and upwelling. Ecol Lett 7:31–41. doi: 10.1046/j.1461-0248.2003.00542.x

[7] Menge BA, Lubchenco J, Bracken MES, Chan F, Foley MM, Freidenburg TL, Gaines SD, Hudson G, Krenz C, Leslie H, Menge DNL, Russell R, Webster MS (2003) Coastal

oceanography sets the pace of rocky intertidal community dynamics. Proc Natl Acad Sci USA 100 (21):12229–12234. doi: 10.1073/pnas.1534875100

[8] Navarrete SA, Wieters EA, Broitman BR, Castilla JC (2005) Scales of benthic–pelagic coupling and the intensity of species interactions: from recruitment limitation to top down control. Proc Natl Acad Sci USA 102:18046–18051

[9] Thorson G (1950) Reproductive and larval ecology of marine bottom invertebrates. Biol Rev 25:1–45. doi: 10.1111/j.1469-185X.1950.tb00585.x

[10] Giménez L (2006) Phenotypic links in complex life cycles: conclusions from studies with decapod crustaceans. Integr Comp Biol 46:615–622. doi: 10.1093/icb/icl010

[11] Shima J, Swearer S (2009) Larval quality is shaped by matrix effects: implications for connectivity in a marine metapopulation. Ecology 90:1255–1267. doi: 10.1890/08-0029.1

[12] Basch L, Pearse J (1995) Consequences of larval feeding environment for settlement and metamorphosis of a temperate echinoderm. Oceanol Acta 19:273–285.

[13] Pechenik J, Estrella M, Hammer K (1996) Food limitation stimulates metamorphosis of competent larvae and alters postmetamorphic growth rate in the marine prosobranch gastropod *Crepidula fornicata*. Mar Biol 127:267–275. doi: 10.1007/BF00942112

[14] Anil A, Desai D, Khandeparker L (2001) Larval development and metamorphosis in *Balanus amphitrite* Darwin (Cirripedia; Thoracica): significance of food concentration, temperature and nucleic acids. J Exp Mar Bio Ecol 263:125–141. doi: 10.1016/S0022-0981(01)00280-5

[15] Pechenik J, Tyrell A (2015) Larval diet alters larval growth rates and post-metamorphic performance in the marine gastropod *Crepidula fornicata*. Mar Biol 162(8):1597-1610. doi: 10.1007/s00227-015-2696-7

[16] Pechenik J, Jarrett J, Rooney J (2002) Relationships between larval nutritional experience, larval growth rates, juvenile growth rates, and juvenile feeding rates in the prosobranch gastropod *Crepidula fornicata*. J Exp Mar Bio Ecol 280:63–78. doi: 10.1016/S0022-0981(02)00367-2

[17] Menge B (1992) Community regulation: under what conditions are bottum-up factors important on rocky shores? Ecology 73:755–765. doi: 10.2307/1940155

[18] Gaines S, Hudson G, Krenz C, et al (2003) Coastal oceanography sets the pace of rocky intertidal community dynamics. Proc Natl Acad Sci USA 100:12229–12234. doi: 10.1073/pnas.1534875100

[19] Leslie H, Breck E, Chan F, et al (2005) Barnacle reproductive hotspots linked to nearshore ocean conditions. Proc Natl Acad Sci USA 102:10534–10539. doi: 10.1073/pnas.0503874102

[20] Burrows MT, Jenkins SR, Robb L, Harvey R (2010) Spatial variation in size and density of adult and post-settlement *Semibalanus balanoides*: effects of oceanographic and local conditions. Mar Ecol Prog Ser 398:207–219. doi: 10.3354/meps08340

[21] Mazzuco, ACA, Pineda, J, Starczak, VC, Christofoletti, RA, Ciotti, AM (2015) Temporal variation in intertidal community recruitment and its relationships to physical forcings, chlorophyll-a concentration and sea surface temperature. Mar Biol 162:1705–1725. doi: 10.1007/s00227-015-2689-6

[22] Menge B, Daley B, Wheeler P, Strub P (1997) Rocky intertidal oceanography: an association between community structure and nearshore phytoplankton concentration. Limnol Oceanogr 42:57–66. doi: 10.4319/lo.1997.42.1.0057

[23] Page HM, Hubbard DM (1987) Temporal and spatial patterns of growth in mussels *Mytilus edulis* on an offshore platform: relationships to water temperature and food availability. J Exp Mar Bio Ecol 111:159–179. doi: 10.1016/0022-0981(87)90053-0

[24] Sanford E, Menge B (2001) Spatial and temporal variation in barnacle growth in a coastal upwelling system. Mar Ecol Prog Ser 209:143–157. doi: 10.3354/meps209143

[25] Figueiras FG, Labarta U, Fernández Reiriz MJ (2002) Coastal upwelling, primary production and mussel growth in the Rías Baixas of Galicia. Hydrobiologia 484:121–131. doi: 10.1023/A

[26] Stone C (1989) A comparison of algal diets for cirripede nauplii. J Exp Mar Bio Ecol 132: 17–40. https://doi.org/10.1016/0022-0981(89)90174-3

[27] Olson R, Olson M (1989) Food limitation of planktotrophic marine invertebrate larvae: does it control recruitment success? Annu Rev Ecol Syst 20:225-247.

[28] Starr M, Himmelman J, Therriault J (1991) Coupling of nauplii release in barnacles with phytoplankton blooms: a parallel strategy to that of spawning in urchins and mussels. J Plankton Res 13:561–571. doi: 10.1093/plankt/13.3.561

[29] Bueno M, Moser G, Tocci B, Flores A (2010) Retention-favorable timing of propagule release in barnacles and periwinkles. Mar Ecol Prog Ser 414:155–165. doi: 10.3354/meps 08715

[30] Kasten P, Flores A (2013) Disruption of endogenous tidal rhythms of larval release linked to food supply and heat stress in an intertidal barnacle. Mar Ecol Prog Ser 472:185–198. doi: 10.3354/meps10005

[31] Hines AH (1979) The comparative reproduction ecology of three species of intertidal barnacles. In: Stancyk S.E., ed. Reproductive ecology of marine invertebrates. pp. 213–234. Belle Baruch Library In Marine Sciences, Univ. S. Carolina Press, Columbia.

[32] Kautsky N (1982) Quantitative studies on gonad cycle, fecundity, reproductive output and recruitment in a baltic *Mytilus edulis* population. Mar Biol 68:143–160. doi: 10.1007/BF00397601

[33] Barnes H (1962) Note on variations in the release of nauplii of *Balanus balanoides* with special reference to the spring diatom outburst. Crustaceana 4:118–122.

[34] Desai D, Anil A (2005) Recruitment of the barnacle *Balanus amphitrite* in a tropical estuary: implications of environmental perturbation, reproduction and larval ecology. J Mar Biol Assoc UK 85:909–920. doi: 10.1017/S0025315405011884

[35] Barbosa A, Gomes C, Pereira G, et al (2016) Local biological drivers, not remote forcing, predict settlement rate to a subtropical barnacle population. Mar Ecol Prog Ser 543:201–208. doi: 10.3354/meps11589

[36] Toupoint N, Gilmore-Solomon L, Bourque F, et al (2012) Match/mismatch between the *Mytilus edulis* larval supply and seston quality: effect on recruitment. Ecology 93:1922–1934. doi: 10.1890/11-1292.1

[37] Pineda J, Reyns N, Starczak V (2009) Complexity and simplification in understanding recruitment in benthic populations. Popul Ecol 51:17–32. doi: 10.1007/s10144-008-0118-0

[38] Bernardo J (1996) Maternal effects in animal ecology. Integr Comp Biol 36 (2): 83–105. https://doi.org/10.1093/icb/36.2.83

[39] Marshall J, Uller T (2007) When is a maternal effect adaptive? Oikos 116:1957–1963. doi: 10.1111/j.2007.0030-1299.16203.x

[40] McEdward LR (1995), Jaeckle WB (1995) Variation in the size, energy content, and biochemical composition of invertebrate eggs: correlates to the mode of larval development. In: McEdward L (ed) Ecology of marine invertebrate larvae. CRC Press, Boca Raton, FL, p 49–77.

[41] Freuchet F, Tremblay R, Flores A (2015) Interacting environmental stressors modulate reproductive output and larval performance in a tropical intertidal barnacle. Mar Ecol Prog Ser 532:161–175. doi: 10.3354/meps11377

[42] McCann M, Padilla D (2015) Effects of a patchy food environment across life history stages. J Exp Mar Bio Ecol 472:135–141. doi: 10.1016/j.jembe.2015.07.009

[43] Pernet F, Tremblay R, Langdon C, Bourget E (2004) Effect of additions of dietary triacylglycerol microspheres on growth, survival, and settlement of mussel (Mytilus sp.) larvae. Mar Biol 144:693–703. doi: 10.1007/s00227-003-1234-1

[44] Gagné R, Tremblay R, Pernet F, et al (2010) Lipid requirements of the scallop *Pecten maximus* (L.) during larval and post-larval development in relation to addition of *Rhodomonas salina* in diet. Aquaculture 309:212–221. doi: 10.1016/j.aquaculture.2010.09.040

[45] Qiu J, Gosselin L, Qian P (1997) Effects of short-term variation in food availability on larval development in the barnacle *Balanus amphitrite amphitrite*. Mar Ecol Prog Ser 161:83–91. doi: 10.3354/meps161083

[46] Thiyagarajan V, Harder T, Qian P (2002) Relationship between cyprid energy reserves and metamorphosis in the barnacle *Balanus amphitrite* Darwin (Cirripedia; Thoracica). J Exp Mar Bio Ecol 280:79–93. doi: 10.1016/S0022-0981(02)00415-X

[47] Pechenik J, Wendt D, Jarrett J (1998) Metamorphosis is not a new beginning larval experience influences juvenile performance. Bioscience 48:901–910. doi: 10.2307/1313294

[48] Harder T, Thiyagarajan V, Qian P (2001) Combined effect of cyprid age and lipid content on larval attachment and metamorphosis of *Balanus amphitrite* Darwin. Biofouling 17:257–262. doi: 10.1080/08927010109378486

[49] Thiyagarajan V, Harder T, Qiu J, Qian P (2003) Energy content at metamorphosis and growth rate of the early juvenile barnacle *Balanus amphitrite*. Mar Biol 143:543–554. doi: 10.1007/s00227-003-1077-9

[50] Pechenik J (2006) Larval experience and latent effects–metamorphosis is not a new beginning. Integr Comp Biol 46:323–333. doi: 10.1093/icb/icj028

[51] Chiu J, Wang H, Thiyagarajan V, Qian P (2008) Differential timing of larval starvation effects on filtration rate and growth in juvenile *Crepidula onyx*. Mar Biol 154:91–98. doi: 10.1007/s00227-007-0902-y

[52] Cullen JJ, Franks PJS, Karl DM, Longhurst A (2002) Physical influences on marine ecosystem dynamics. In: Robinson A.R., McCarthy J.J. and Rothchild B.J. (eds.), The Sea - Volume 12. pp. 297–336. Wiley, New York.

[53] Eckman JE, Wener FE, Gross TF (1994) Modeling some effects of behavior on larval settlment in a turbulent boundary-layer. Deep Sea Res Part II Trop Stud Oceanogr 41 (1):185–208. doi: 10.1016/0967-0645(94)90067-1

[54] Gaylord B, Hodin J, Ferner MC (2013) Turbulent shear spurs settlement in larval sea urchins. Proc Natl Acad Sci USA 110 (17): 6901–6906. doi: 10.1073/pnas.1220680110

[55] Wheeler JD, Chan KYK, Anderson EJ, Mullineaux LS (2016) Ontogenetic changes in larval swimming and orientation of pre-competent sea urchin *Arbacia punctulata* in turbulence. J Exp Biol 219(9): 1303–1310. doi: 10.1242/jeb.129502

[56] Queiroga H, Cruz T, dos Santos A, Dubert J, Gonzalez-Gordillo JI, Paula J, Peliz A, Santos AMP (2007) Oceanographic and behavioural processes affecting invertebrate larval dispersal and supply in the western Iberia upwelling ecosystem. Prog Oceanogr 74(2–3): 174–191.

[57] Saunders MI, Metaxas A (2010). Physical forcing of distributions of bryozoan cyphonautes larvae in a coastal embayment. Mar Ecol Prog Ser 418:131–145. doi:10.3354/meps08842

[58] Tapia FJ, DiBacco C, Jarrett J, Pineda J (2010) Vertical distribution of barnacle larvae at a fixed nearshore station in southern California: stage-specific and diel patterns. Estuar Coast Shelf Sci 86 (2): 265–270. doi: 10.1016/j.ecss.2009.11.003

[59] Queiroga H, Blanton J (2005) Interactions between behavior and physical forcing in the control of horizontal transport of decapods crustacean larvae. Adv Mar Biol 47:107–214. doi: 10.1016/S0065-2881(04)47002-3

[60] Graham WM, Field JG, Potts DC (2013) Persistent upwelling shadows and their influence on zooplankton distributions. Mar Biol 114 (4): 561–570. doi: 10.1007/BF00357253

[61] Morgan SG, Fisher JL (2010) Larval behavior regulates nearshore retention and offshore migration in an upwelling shadow and along the open coast. Mar Ecol Prog Ser 404: 109–126. doi: 10.3354/meps08476

[62] Farfan LM, D'Sa EJ, Liu KB, Rivera-Monroy VH (2014) Tropical cyclone impacts on coastal regions: the case of the Yucatan and the Baja California Peninsulas, Mexico. Estuar Coast 37(6):1388–1402. doi: 10.1007/s12237-014-9797-2

[63] Gyory J, Pineda J (2011) High-frequency observations of early-stage larval abundance: do storms trigger synchronous larval release in *Semibalanus balanoides*? Mar Biol 158(7): 1581–1589. doi: 10.1007/s00227-011-1671-1

[64] Sutherland JP (1990) Recruitment regulates demographic variation in a tropical intertidal barnacle. Ecology 71:955–972. doi: 10.2307/1937365

[65] Thompson RC, Norton TA, Hawkins SJ (1998) The influence of epilithic microbial films on the settlement of *Semibalanus balanoides* cyprids – a comparison between laboratory and field experiments. Hydrobiologia 375: 203–216. doi: 10.1023/A:1017036301082

[66] Yang J, Li X, Liang X, Bao W, Shen H, Li J (2014) Effects of natural biofilms on settlement of plantigrades of the mussel *Mytilus coruscus*. Aquaculture 424: 228–233. doi: 10.1016/j.aquaculture.2014.01.007

[67] Pineda J, Starczak V, Stueckle T (2006) Timing of successful settlement: demonstration of a recruitment window in the barnacle *Semibalanus balanoides*. Mar Ecol Prog Ser 320:233–237. doi: 10.3354/meps320233

[68] Vargas CA, Manríquez PH, Navarrete SA (2006) Feeding by larvae of intertidal invertebrates: assessing their position in pelagic food webs. Ecology 87(2):444–457. doi: 10.1890/05-0265

[69] Krenz C, Menge BA, Freidenburg TL, Lubchenco J, Chan F, Foley MM, Nielsen KJ (2011) Ecological subsidies to rocky intertidal communities: linear or non-linear changes along a consistent geographic upwelling transition? J Exp Mar Bio Ecol 409 (1–2): 361–370. doi: 10.1016/j.jembe.2011.10.003

[70] Pfaff MC, Branch GM, Wieters EA, Branch RA, Broitman BR (2011) Upwelling intensity and wave exposure determine recruitment of intertidal mussels and barnacles in the southern Benguela upwelling region. Mar Ecol Prog Ser 425:141–152. doi: 10.3354/meps09003

[71] Reddin CJ, Docmac F, O'Connor NE, Bothwell JH, Harrod C (2015) Coastal upwelling drives intertidal assemblage structure and trophic ecology. Plos One 10(7): e0130789. doi: 10.1371/journal.pone.0130789

[72] Menge BA, Blanchette C, Raimondi P, Freidenburg T, Gaines S, Lubchenco J, Lohse D, Hudson G, Foley M, Pamplin J (2004) Species interaction strength: testing model predictions along an upwelling gradient. Ecol Monogr 74(4): 663–684. doi: 10.1890/03-4060

[73] Shanks AL, Mcculloch A, Miller J (2003) Topographically generated fronts, very nearshore oceanography and the distribution of larval invertebrates and holoplankters. J Plank Res 25(10):1251–1277. doi: 10.1093/plankt/fbg090

[74] Menge BA, Gouhier TC, Freidenburg T, Lubchenco J (2011). Linking long-term, large-scale climatic and environmental variability to patterns of marine invertebrate recruitment: toward explaining "unexplained" variation. J Exp Mar Biol Ecol, 400(1–2): 236–249. doi: 10.1016/j.jembe.2011.02.003

[75] Blanchette CA, Broitman BR, Gaines SD (2006). Intertidal community structure and oceanographic patterns around Santa Cruz Island, CA, USA. Mar Biol, 149: 689–701. doi: 10.1007/s00227-005-0239-3

[76] Byrne M, Przeslawski R. (2013) Multistressor impacts of warming and acidification of the ocean on marine invertebrates' life histories. Integr Comp Biol 53(4): 582–596. doi: 10.1093/icb/ict049

[77] Menge BA, Chan F, Nielsen KJ, Di Lorenzo E, Lubchenco J (2009). Climatic variation alters supply-side ecology: impact of climate patterns on phytoplankton and mussel recruitment. Ecol Monogr, 79(3): 379–395. doi: 10.1890/08-2086.1

Chapter 3

How Does Chloroplast Protect Chlorophyll Against Excessive Light?

Lucia Guidi, Massimiliano Tattini and Marco Landi

Additional information is available at the end of the chapter

http://dx.doi.org/10.5772/67887

Abstract

Chlorophylls (Chls) are the most abundant plant pigments on Earth. Chls are located in the membrane of thylakoids where they constitute the two photosystems (PSII and PSI) of terrestrial plants, responsible for both light absorption and transduction of chemical energy via photosynthesis. The high efficiency of photosystems in terms of light absorption correlates with the need to protect themselves against absorption of excess light, a process that leads to the so-called photoinhibition. Dynamic photoinhibition consists of the downregulation of photosynthesis quantum yield and a series of photo-protective mechanisms aimed to reduce the amount of light reaching the chloroplast and/or to counteract the production of reactive oxygen species (ROS) that can be grouped in: (i) the first line of chloroplast defence: non-photochemical quenching (NPQ), that is, the dissipation of excess excitation light as heat, a process that takes place in the external antennae of PSII and in which other pigments, that is carotenoids, are directly involved; (ii) the second line of defence: enzymatic antioxidant and antioxidant molecules that scavenge the generated ROS; alternative electron transport (cyclic electron transport, pseudo-cyclic electron flow, chlororespiration and water-water cycle) can efficiently prevent the over-reduction of electron flow, and reduced ferredoxin (Fd) plays a key role in this context.

Keywords: antioxidant, carotenoids, excess excitation energy, non-photochemical quenching, photosystem

1. Introduction

Pigments in plants, cyanobacteria, algae and photosynthetic anoxygenic bacteria are the most important molecules involved in photosynthesis, the only biological process that tunnels energy on Earth. Pigments play two key roles in photosynthesis: they absorb sunlight and transduce it into chemical energy. The most important pigment is certainly chlorophyll (Chl), an organic compound that typically shows chlorine, a cyclic tetrapyrrole ring, coordinated to a central atom of magnesium (**Figure 1**). This molecular structure is very similar to that found in the eme group in which the central atom is iron. Diversification of various Chls is due to the different side chains bonded to the chlorine ring (Chl *a*, *b*, *c*, *d*, *e* and *f*).

The process of light absorption consists of a sequence of photophysical and photochemical reactions that are subdivided into three stages: (i) light absorption, (ii) utilization of this energy to synthesize ATP and reducing power, reduced ferredoxin (Fd) and NADPH and (iii) absorption and reduction of atmospheric CO_2 into carbon skeleton. However, the most important and true light reaction is represented by charge separation that occurs at the reaction

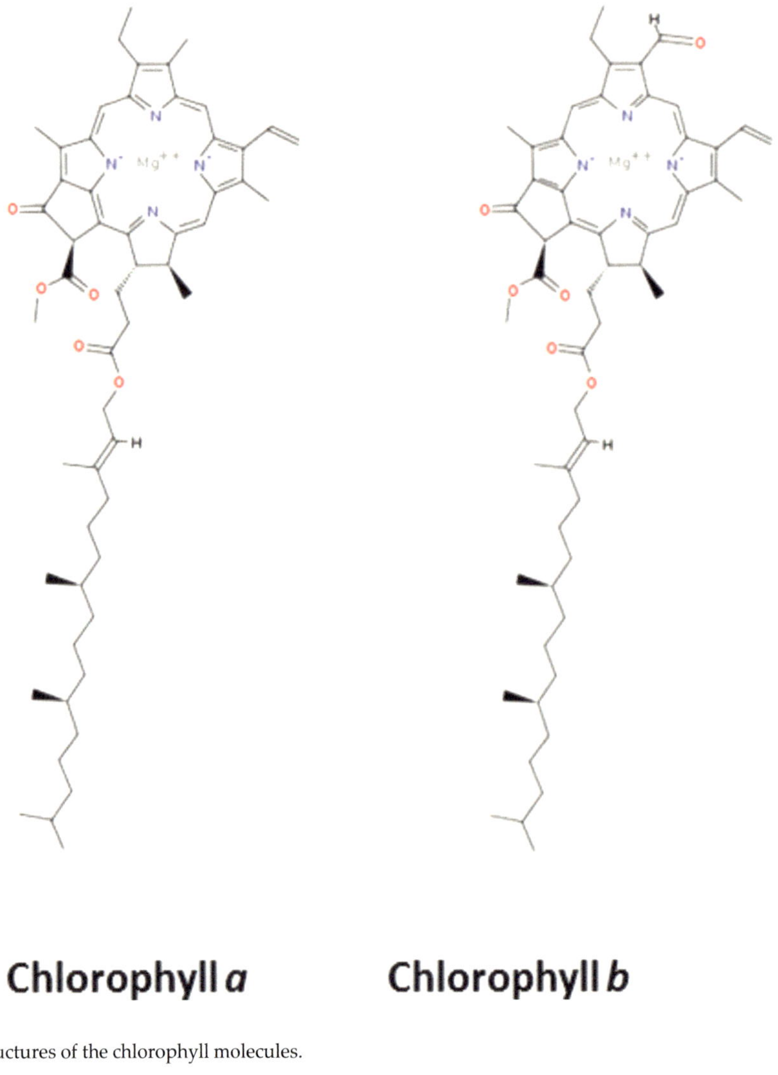

Figure 1. Structures of the chlorophyll molecules.

http://dx.doi.org/10.5772/67887

centres. The process is possible for the presence of organic molecules able to capture sunlight and transduce it in chemical energy namely photosynthetic pigments and that is chlorophylls and, carotenoids. These pigments aggregate with proteins and act as an antenna harvesting the energy of sunlight and tunnelling this energy into the reaction centres located in photosystems. In plants and algae, there are about 200–400 light harvesting molecules. Light harvesting complexes have evolved many adaptive mechanisms that permit photosynthetic organisms to thrive in different environments. The spectral distribution of sunlight that reaches our planet largely covers the absorption spectra of photosynthetic pigments utilized in light harvesting antennas (**Figure 2**). In a general way, light harvesting antennas have developed the ability to optimize light capture under both low- and high-intensity light conditions [1].

The optimal absorption wavelength range for light harvesting antennas is in the red region (680–690 nm), where the energy is utilized by chlorophyll to split water and reduce ferredoxin. The evolution of the most abundant pigments, chlorophyll *a*, is probably related to its efficient absorption in this region in addition to, perhaps, its chemistry and for its redox potential.

All photosynthetic pigments show a chromophore, which possesses two orbitals whose difference in energy falls within the light spectrum. In consequence, a photon of incident light is able to excite an electron from its ground-state orbital to the excited state. From a chemical point of view, the chromophore exists as conjugated π-electron systems or metal complexes. In a conjugated π system, electron excitation occurs between π orbitals spread across alternating single and double bounds (e.g., carotenoids). The metal complex chromophores share d orbitals between transition metals and ligands (e.g., chlorophylls). Really, in the antenna pigments, chromophores are not individual entities, and they synergically interact with each other and this interaction plays a crucial role in the light harvesting mechanism.

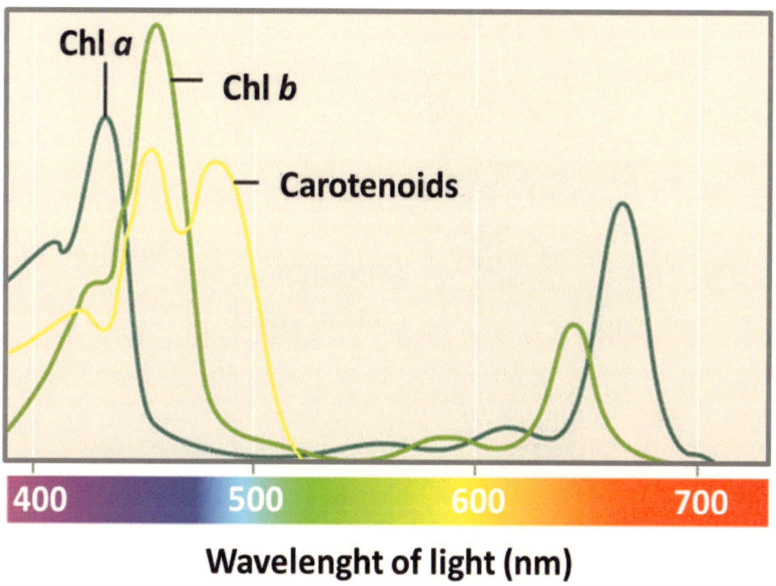

Figure 2. Chlorophyll *a*, *b* and carotenoids absorbance spectra.

Light-harvesting complex (LHC) is the complex of subunit proteins that may be part of a larger supercomplex of a photosystem and is the functional unit in photosynthesis, devoted to the absorption of sunlight. The energy excitation is first tunnelled among other surrounding molecules of the same complex and then from one LHC to another and then funnelled to reaction centres (RCs), where it is converted into charge separation with 90% quantum efficiency.

The presence of proteins in LHC complexes is attributable to the fact that Chl of RCs cannot absorb sunlight at an efficient rate that is enough for efficient photosynthesis to occur. In fact, Chl molecules in RCs absorb only a few photons each second, which are insufficient to drive electron transport into chloroplast membranes (present in 1 RC of about 300 antenna molecules). To overcome this problem, RCs are associated with antenna pigment-protein complexes that absorb sunlight and very efficiently transfer it to RCs. For the importance of the LHCs in gathering sunlight, they differ in the number of pigments and in their composition and structure in a way that they are an optimized energy collector system (**Figure 3**). The proteins play an important function in the precise position, mutual separation and relative orientation of antenna.

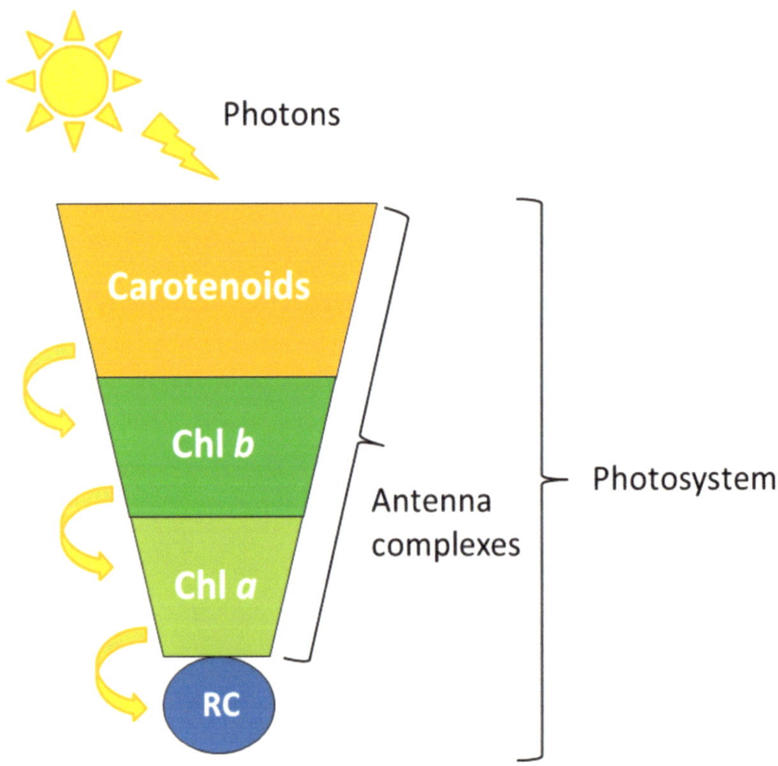

Figure 3. A schematic representation of the light absorption process of chloroplasts. Antenna complexes, composed of carotenoids, Chl *a* and Chl *b* molecules, absorb photons from sunlight and transfer them to the RC, which consists of a special couple of molecules of Chl *a*. Antenna complexes and the RC form a photosystem.

http://dx.doi.org/10.5772/67887

Photosynthetic unit (PSU) represents the basic unit of the light-harvesting apparatus and consists of a large number of antenna chromophores coupled to a RC. Excitation-transfer pathways follow a scheme in which different chromophores build an energy funnel where chromophores, which absorb in the blue side of spectrum, transfer excitation energy to more red-shifted chromophores (**Figure 3**). Theoretically, the PSUs are considered individual entities but [2] proposed the *lake* and *puddle* model. In the second model, the PSUs do not interact with each other and the excitation light absorbed by chromophores is always transferred to the same RC. Differently, in the *lake* model, the antenna chromophores form a matrix with embedded RCs in which there is an unrestricted energy transfer.

2. Charge separation in photosystems and electron transport

Photosynthesis starts with light absorption by the chromophores, which excites the molecules from the ground state to an electronic excited state. Once sunlight energy is absorbed, pigments in the excited state have a short life and relax to the ground state after about 4 ns [3]. The singlet excited state lifetime of Chl is lower compared with the radiative lifetime, largely owing to intersystem crossing, which yields triplet excited states of Chl (about 10 ns) [4]. This electronic excitation must be usefully harvested before the molecules relax, and this happens when excitons are transferred through space among chromophores until they reach, eventually, a RC where charge separation occurs. In plants, there are two RCs constituted by two Chl molecules, P680 and P700, respectively, for PSII and PSI, and Chl with absorbance maxima corresponding to these wavelengths is proposed as the final slight sink. These chlorophylls drive electron transfer by charge separation, a reaction in which P680 and P700 molecules reduce an acceptor. These driving reactions energetically downhill from the potential that is more negative to ones that are more positive (**Figure 4**). All these electron transfer steps in photosynthesis share a common feature. The loss of an electron from one component, which remains in an oxidized state, reduces another one. Typically, electron transport carriers are small molecules or atoms of metallic elements that can exist in a number of valence states.

In photosystem RCs, the light-induced loss of an electron (charge separation) leaves P680 and P700 in an oxidized state ($P680^+$ and $P700^+$) and the respective acceptors, pheophytin for P680 and A_0 (chlorophyll), in a reduced state. $P680^+$ is reduced from an adjacent tyrosine molecule (TyrZ) in the polypeptide chain of the D1 protein of the PSII complex. In turn, the oxidized is reduced by electrons from the oxygen-evolution complex (OEC) that oxidized water. Two water molecules are oxidized to produce oxygen, four protons and four electrons that are transferred one at a time. These redox reactions are carried out by OEC that consists of four manganese atoms held in a protein matrix with one atom of calcium and chlorine each (**Figure 4**). This process is known as a S-cycle from [5] that provides protons derived from water oxidation to be released into the lumen of the thylakoid membranes.

In the other set of reactions, reduced pheophytin is oxidized by passing an electron to the first of two plastoquinone (PQ) molecules, tightly bound at the site Q_A of D2 protein in the PSII. Then, via an iron atom, an electron is transferred to the next PQ at the site Q_B. Both PQs require two electrons for their complete reduction; at the Q_A site, PQ undergoes to a single

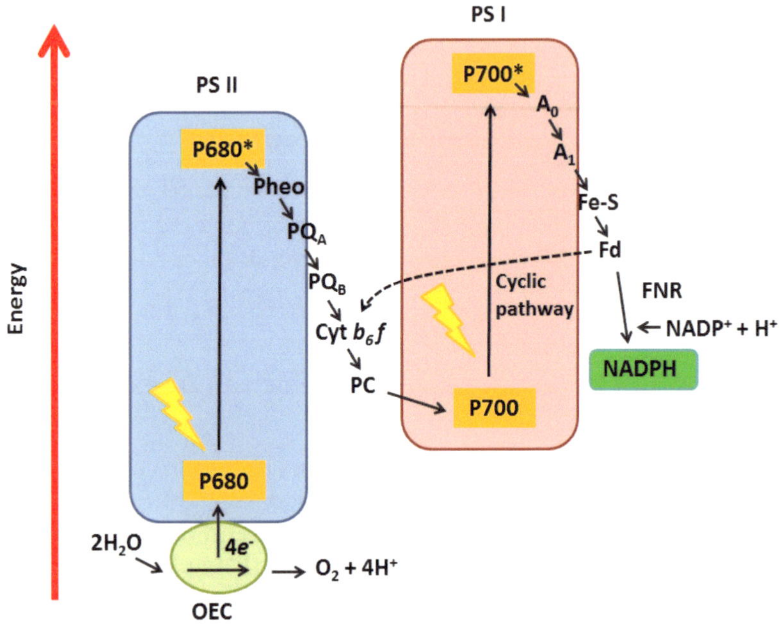

Figure 4. A representation of the linear non-cyclic (solid line) and cyclic electron flow (dashed line) in the chloroplast membranes. OEC tetranuclear Mn cluster; P680, reaction centre of photosystem II (PSII); P680*, excited electronic state of P680; Ph, pheophytin; Q_A and Q_B, plastoquinone; protein complex containing cytochrome b_6 and cytochrome *f*; PC, plastocyanin; P700, reaction centre of PSI; P700*, excited electronic state of P700; A_0, a special chlorophyll *a* molecule; A_1, phylloquinone; Fe-S, iron sulphur centres; Fd, ferredoxin; NADP, nicotinamide-adenine dinucleotide phosphate and FNR, ferredoxin-NADP$^+$ reductase.

reduction event to the semiquinone state before being re-oxidized by the PQ at Q_B site. Two successive reductions occur that fully reduce PQ at Q_B site, which, for its reduction, requires also two protons from the stromal side of the membranes and forms PQH_2 that leaves PSII and diffuses in the lipid bilayer, representing a mobile carrier of protons and electrons. A new molecule of PQ (in oxidized form) replaces this plastoquinone in the Q_B site.

PQH_2 formed by the PSI activity represents the substrate of the Q cycle on cytochrome b_6f, another integral transmembrane protein complex on thylakoid membranes. PQH_2 is oxidized in two steps to PQ. The first step happens at Q_p site, located on the luminal side of cytochrome b_6f, and the electron is transferred at the end to plastocyanin (PC), a soluble small protein containing copper. The second electron is transferred until Q_n site located on the stromal side of the cytochrome where it reduces further PQ molecule to semiplastoquinone. Another PQH_2 molecule originating from PSII is oxidized in the same two steps at the Q_p site, generating further a reduced plastocyanin and completing the reduction of semiplastoquinone to PQH_2. The oxidation of PQH_2 at Q_p site determines the release of two protons in the lumen that represents the most important feature of the Q cycle. In fact, this cycle acts as a proton pump, essential to generate the transmembrane electrochemical H^+ gradient.

After light absorption and charge separation in PSI, P700$^+$ is generated, and it is reduced back to P700 by direct interaction with reduced PC diffusing from cytochrome b_6f complex.

http://dx.doi.org/10.5772/67887

Plastocyanin, from its copper atom, reduces directly P700$^+$. The electron flow generated by charge separation that occurs in P700 determines the reduction of different carriers, and the final electron acceptor is represented by Fd, a small water-soluble iron-sulphur protein. Reduced Fd is capable of reducing a variety of molecules. Usually, it reduces NADP$^+$, which requires two electrons and two protons to yield NADPH in a reaction catalyzed by ferredoxin-NADP$^+$ reductase (FNR) (**Figure 4**), thus completing the so-called Z scheme. The electron flow generates even chemical energy, that is ATP, by the enzymatic activity of ATP-ase, a transmembrane complex that utilizes the proton gradient generated by Q cycle and water oxidation, to synthetize ATP.

3. Excess of excitation energy

In the past, the higher order structure of PSII was thought to be important only to increase the efficiency of light harvesting; nowadays, it has been suggested that it provides the essential dynamic properties involved in its regulation [6]. When light is low, in a way, extremely efficient antenna systems absorb light and tunnel it through RC, but when light is in excess, a large extent of this energy is dissipated, overall as heat, to prevent photo-damage to PSUs. When plants are exposed to shade or sunlight conditions, different mechanisms occur. Shade leaves are typically larger in area but thinner than sun leaves because they develop shorter palisade cells. In shade leaves, the chloroplasts move within the cells to take up a position where they will absorb the maximum light without shading other chloroplasts below. In addition, shade leaves show a large number of antenna, and usually, the peripheral antenna are rich in Chl *b* molecules (Chl *a*/*b* = 1.33). All these mechanisms enhance and optimize the light absorption. However, even shade leaves have adapted mechanisms aimed to regulate the light absorption, as the state II-I transition (also called spillover process). The aim of this process is the reduction of light tunnelled to P680 altering the ratio of light energy absorbed between PSII and PSI. In fact, RCs of the two photosystems have different absorption spectra (high energy is absorbed by P680 as compared with P700), and this determines that when the energy flow through each is not balanced to the requirement of the Z scheme, an excess of energy could accumulate in the system. In this way, LHCII trimers represent a feedback loop that adjusts the amount of antenna Chls, providing energy to each photosystem (state transition). The excess of light energy flowing through PSII RCs is higher than that flowing through PSI RCs, conditions in which an excess of reduced PQ occurs. This activates a kinase that phosphorylates some LHCII trimers, and this extra charge allows them to dissociate from the PSII (state II) and migrate towards the stroma lamellae (state I transition) where they bind to the PSI complex, increasing in this way the flow through the system. The increase of PSI activity leads to the oxidation of reduced PQ, which activates a phosphatase that removes the phosphate group to the LHCII trimers that return to PSII (state II transition).

In contrast, sun leaves live in very high radiation levels overall at the top of the canopy. The light response curve in relation to the light intensity shows that the amount of energy utilized is lower than that absorbed because the light energy utilized in carbon reduction is mostly due to the limitation on the rate of CO_2 diffusing into the leaf (**Figure 5**). In these conditions,

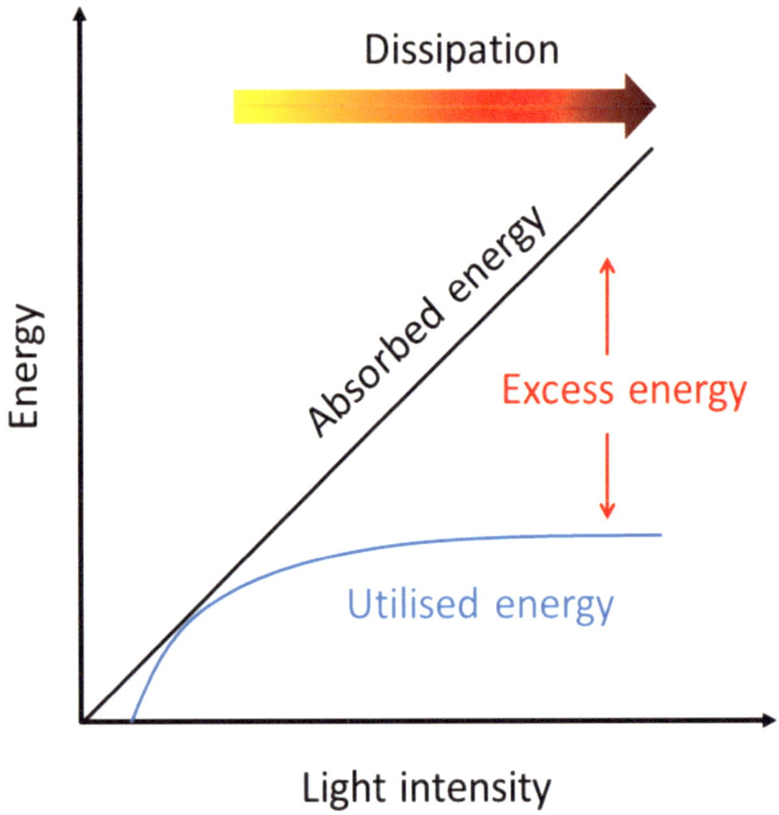

Figure 5. Absorbed and utilized energy in response to increasing light intensities. When light absorbed exceed photosystems requirement, the 'excess energy' can potentially cause photo-oxidative damage if it is not efficiently dissipated.

the antenna Chls become saturated and tunnel a high flow of the excitation energy to the RC that cannot be dissipated along the electron flow. The excess of energy must be efficiently dissipated through different mechanisms in order to avoid photo-damage to PSII.

Photosystem II is particularly sensitive to photoinhibition because the high redox potential of the oxidized P680 ($P680^+$), on the other hand, necessary for water oxidation. Accumulation of $P680^+$ leads to different types of photoinhibition:

(i) Acceptor-side photoinhibition: when reduced PQ is not re-oxidized, the $P680^*$ charge recombination is inhibited and P680 is expected to lead to the triplet state of P680, $^{T}P680^*$. This chemical species may react with oxygen and produce harmful singlet oxygen.

(ii) Donor-side photoinhibition: if the OEC is chemically inactivated, the donation of electrons from water does not keep up with the electron transfer from P680 to the acceptor side. In this case, an accumulation of $P680^+$ occurs. The high redox potential of this chemical species induces the oxidation of various organic components such as proteins or pigments until damage is done to D1 protein of PSII.

http://dx.doi.org/10.5772/67887

Different mechanisms are present in PSII aimed to dissipate the excess of photons absorbed by antenna, and different defence lines occur into the chloroplast.

4. First line of defence of chloroplast: dissipation of excess excitation light

First line of chloroplast defence includes suppression mechanisms aimed to reduce or dissipate the excitation light tunnelled in P680. At leaf level, the change in the leaf angle with respect to the incident light and/or the chloroplast movement into the leaf to self-shading positions along the sidewalls of cells represent mechanisms by which a decrease in absorbed light can occur.

In the chloroplast, there are essentially three mechanisms to contrast the high light conditions: adjustment in synthesis and amount of antenna protein, movement of LHCII (state II-I transition) and non-photochemical quenching [7]. The first of these mechanisms is related to the expression of *Lhcb* genes, whose expression is downregulated by high light conditions and/or low CO_2 concentration. The sensor mechanism is not known even though one possible candidate is the redox potential (i.e., the level of reduced PQ) [8], but also ROS represent possible signal molecules [9, 10]. Clearly, these slow mechanisms cannot entirely prevent the accumulation of excess of energy in the antenna system. However, photosynthesis in green plants depends on protective mechanisms that adapt within minutes or seconds to changing light conditions. Excited Chls return to the ground state either by emitting photons (fluorescence) or by dissipating it as heat. All these mechanisms aimed to remove this trapped energy before it passed on down the electron transport chain are named *non-photochemical quenching* (NPQ). NPQ is heterogeneous and composed by at least three components: the major and rapid component is the pH- or energy-dependent component qE, a second component qT, related to the phenomenon of state transition but negligible in most of plants under excess light and the third and slow component, qI, related to the photoinhibition of photosynthesis [11].

It has been reported that two distinct qE mechanisms occur, one involving zeaxanthin (Zea) (quenching type 1) and the other carotenoid lutein (Lut) (quenching type 2) [12]. In qE type I, three xanthophylls, violaxanthin (Vio), anteraxanthin (Ant) and Zea, are involved in the well-known xanthophyll cycle in which the epoxidation of Vio to Zea via Ant determines an efficient dissipation of excess light into heat [13]. Electron flow pumping and generating protons in the lumen decrease its pH from about 7 to less than 5; this represents a strong signal that starts a series of quenching processes. The low pH-induced protonation of PsbS peptide, for its proximity to antenna complexes (CP24, CP26 and CP29), induces in turn in these complexes conformational changes. In the chemical state, antenna complexes bind one molecule of Zea and one of Chl (*Zea-Chl complex = quenching complex*) that accept energy transfer from excited Chls. Zeaxanthins are able to return to their ground state dissipating energy as heat :

$$\text{LHCII}^* + \text{zeaxanthin} \rightarrow \text{LHCII} + \text{zeaxanthin}^*. \quad (1)$$

$$\text{Zeaxanthin}^* \rightarrow \text{zeaxanthin} + \text{heat}. \quad (2)$$

It has been reported that in the crystal structure of LHCII is present Vio, and its peripheral localization suggests that it could be de-epoxidized to Zea by Vio de-epoxidase (VDE), an enzyme that is activated by low lumen pH occurring in high light conditions. The back reaction by Zea epoxidase is slow and causes a sustained quenching that relaxes within 1–3 hours following light stress and depends on the release of Zea from antenna pigments. In conclusion, Zea is certainly considered a regulator of light harvesting for its role in the xanthophyll cycle and carries out three fundamental roles during high light conditions: (i) protection against photo-oxidation due to radical oxygen's attack (because it quenches oxygen singlet energy), (ii) absorption of Chl triplet energy and (iii) absorption of incoming photons and transferring them to neighbouring Chl molecules increasing in this way the overall absorption spectrum of the PSs [14]. In addition, it has been reported that this xanthophyll exhibits an antioxidant function in the thylakoid membrane [15].

In addition, trimeric LHCII binds other types of xanthophylls: two all-*trans*-luteins and a 9-*cis*-noexanthin [16]. The minor monomeric complexes CP24, CP26 and CP29 all bind Lut, and in addition, CP29 binds two xanthophyll cycle carotenoids and one-half to one neoxanthin (Neo), CP24 binds two xanthophyll cycle carotenoids and CP26 binds one xanthophyll cycle carotenoids and one Neo [17, 18]. In the quenching type 2, qE is an intrinsic LCHII property: protein conformational changes alter configurations of bound pigment (normally Lut), which become an efficient quencher of Chl-excited state [12]. A change in the conformational state of another LHCII-bound xanthophyll, Neo, correlates with the extent of quenching. In the model for type 2 quenching proposed by [19], Zea acts not as a quencher but as an allosteric modulator of the ΔpH sensitivity of this intrinsic LHCII quenching process. The two types of quenching involved different xanthophylls that operate at different sites, but there are some similarities in the reasons that both involve ΔpH and PsbS-mediated conformational changes [12].

Given that the xanthophyll cycle quenches only 95% of the triplet Chl [20], the unquenched triplet Chl is the reason for the need of singlet oxygen not only scavenging by carotenoids bound to LHCII but also by carotenoids free in lipid matrix [21]. Lut has the specific property of quenching harmful 3Chl* by binding at site L1 of the major LHCII complex and of other Lhc proteins of plants, thus preventing ROS formation [20]. Neo contributes PSII photoprotection in a dual way: determins conformational change in trimeric LHCII, which reduces light absorption and controls the accessibility of the O_2 to the inner core of the complex [20, 22]. The trimeric organization of LHCII is, definitively, effective in screening the internal protein domain from molecular oxygen [23].

5. Second line of defence of chloroplast: antioxidant enzymes and molecules

As reported above, the excess of excitation energy induces an excess of singlet-excited Chl *a* that is de-excited via thermal dissipation. However, the remaining singlet-excited Chl *a* can convert to triplet-excited Chl that readily reduces molecular oxygen. This determines the synthesis of ROS that is potentially dangerous to organic molecules in the chloroplast. In the second line of defence, antioxidant molecules and enzymes that together scavenge ROS play a key role.

The primary products of molecular oxygen reduction are disproportionate to H_2O_2 and O_2 in a reaction catalyzed by superoxide dismutase (SOD). H_2O_2 produced is then reduced to water with the reducing power of ascorbate (ASA) in a reaction catalyzed by ASA peroxidase (APX), and ASA is oxidized to monodehydroascorbate (MDHA) that is directly reduced to ASA by reduced ferredoxin or NADPH by MDHA reductase. Alternatively, MDHA is spontaneously disproportionated to dehydroascorbate (DHA) and ASA. DHA is then reduced by reduced glutathione (GSH), by the enzyme DHA reductase that produces oxidized glutathione (GSSG) and ASA. Finally, GSSG is reduced again in GSH by the action of GSH reductase, and the reducing power is represented by reduced Fd or NADPH, that, in turn, are reduced by PSI activity. This indicates that any pathway aimed to regenerate ASA utilizes electrons derived from water. For this reason, the previous process is referred as water-water cycle [10].

In addition to the primary antioxidant systems, carotenoids have a protective role against ROS since they are very efficient physical and chemical quenchers of singlet oxygen and potent scavengers of other free radicals [24]. For example, β-carotene, located in the core complex of both PSII and PSI, plays a role as a quencher of Chl triplet and singlet oxygen [25], and the products generated from the oxidation of β-carotene by singlet oxygen represent primary sensor signalling under oxidative stress [26]. Other carotenoids play an important role as antioxidants in the chloroplast. Lut is the most abundant carotenoid in the chloroplast and is required as a quencher [7], while Neo can scavenge superoxide anion [27]. The antioxidant activity of carotenoids is carried out in combination with other lipophilic antioxidants. In this way, it has been reported that Zea, in cooperation with tocopherol, prevented photo-oxidation induced by high light [28], or a strong increase in carotenoids pigment (including those involved in xanthophyll cycle) is reported together with the activity of SOD enzyme following oxidative stress [29]. Again, carotenoids can influence the structure and fluidity of thylakoid membranes [30], that is essential for photosynthetic functions, influence barrier status to ions and oxygen, increase thermostability and protect against lipid peroxidation. In fact, as reported by [30], β-carotene can fluidize the membrane because it can move in the inner hydrophobic part of the membrane, and xanthophyll (and in particular Zea) shows the polar group that orientates these carotenoids perpendicular to the membrane surface.

6. From PSII repair processes to alternative electron sinks

In the last 30–40 years, the susceptibility of D1 protein to photo-damage has been well known, and the concept of the replacement of the damaged D1 protein during the repair cycle of PSII is extensively investigated [13, 31–33]. Moreover, D1 damage has been shown to be directly proportional to light intensity [34].

The repair process of photo-damaged D1 proteins consists of different steps: (i) prompt, partial disassembly of the PSII holocomplex, (ii) exposure of the photo-damaged PSII core to the stroma of the chloroplast, (iii) degradation of photo-damaged D1, (iv) *de novo* D1 biosynthesis and insertion in the thylakoid membrane and (v) re-assembly of the PSII holocomplex, followed by activation of the electron-transport process through the reconstituted D1/D2 heterodimer [35]. The sequence leading to the recovery of photo-damaged PSII is consistent with the frequent D1 turnover in the chloroplast and with the heterogeneity in the configuration and function of PSII.

In the past, the sensibility of PSII was linked to an inherent defect of photosynthetic apparatus but now it is clear how this mechanism of damage-repair of PSII is extremely regulated [33] and protects even PSI from irreversible damage. In fact, the repair mechanisms in PSI are time and high energy consuming, and it has been suggested that the inhibition of PSII is likely to protect PSI [33].

Reduced Fd plays an important role in preventing the over-reduction of electron flow, and a wide range of electron sinks are available in chloroplasts. Electrons are preferentially utilized by the FNR enzyme that produces NADPH for CO_2 photoassimilation or ferredoxin:thioredoxin reductase that synthesizes thioredoxin responsible for the regulation of some enzymes of Calvin-Benson cycle [36]. On the other hand, reduced Fd can release electrons also to ferredoxin:nitrite reductase and sulphite reductase for the reductive assimilation of nitrite [37] and sulphur [38]. Finally, reduced Fd represents an electron donor for fatty acid desaturases [39] and glutamine:oxoglutarate amino transferase [40]. However, when $NADP^+$ is not available, reduced Fd releases its electron to different acceptors whose function is to avoid an over-reduction of PSI [41]. It has been discovered that there is an electron transport driven solely by PSI and scientists called it cyclic electron flow. In this cycle, electrons can be recycled from reduced Fd to PQ and subsequently, to the cytochrome b_6f complex via the Q cycle [42]. Such cyclic flow generates ΔpH and thus ATP without the accumulation of reduced species. In addition, the generated ΔpH may regulate photosynthesis via NPQ (see Section 4). Another electron acceptor of reduced Fd is molecular oxygen inducing the pseudo-cyclic electron flow. The reduction of molecular oxygen with one electron generates superoxide anions in the so-called Mehler reaction, which restores the redox poise when linear electron flow is over-reduced [43]. The radical oxygen species is efficiently removed by water-water cycle. Chlororespiration is another effective electron sink in which reduced Fd is directly involved. In this process, two enzymes play the key role: NADH dehydrogenase complex and nucleus-encoded plastid-localized terminal oxidase (PTOX). The enzyme PTOX catalyzes the reaction in which electrons are transferred from PQH_2 to molecular oxygen forming water [44].

Finally, in addition to the above-reported electron flow, photorespiration is another efficient pathway by which plants adjust the ATP/NADPH ratio and consume the excess of excitation energy.

7. Conclusions

Certainly, Chls represent the key molecules involved in light energy absorption and transduction into chemical energy. Chls absorb the light energy that reaches leaves in a very efficient manner but sometimes, light exceeds photochemistry requirement, and the complexity of photosystems is essential to modulate and dissipate excess of excitation energy. A wide range of responses to environmental stimuli thus characterizes the photoprotection of chloroplasts. The increasing level of complexity from the molecular (pigments and protein) to supramolecular (photosystems) level mirrors the necessity of different time-scale responses (from seconds to months) to modulate light that is (inevitably) absorbed. In the range of seconds to minutes, modulation of the redox state of photosynthetic electron transport activates the non-photochemical quenching of excess of excitation energy not only through xanthophyll cycles [13] but also by II-I state transition [45]. On a larger scale (minutes to hours), modulation of redox state of electron transport induces changes in gene expression (organellar and nucleus) through retrograde regulation that changes the structure of the photosynthetic apparatus [46, 47]. On the time scale from weeks to months, the redox state of electron transport determines changes in plant growth and morphology [48].

Author details

Lucia Guidi[1]*, Massimiliano Tattini[2] and Marco Landi[1]

*Address all correspondence to: lucia.guidi@unipi.it

1 Department of Agriculture, Food and Environment, University of Pisa, Pisa, Italy

2 National Research Council of Italy, Department of Biology, Agriculture and Food Sciences, Institute for Sustainable Plant Protection, Sesto Fiorentino, Florence, Italy

References

[1] Perrine Z, Negi S, Sayre RT. Optimization of photosynthetic light energy utilization by microalgae. Algal Res 2012; **1**: 134–142. doi:10.1016/j.algal.2012.07.002

[2] Bernhardt K, Trissl H-W. Theories for kinetics and yields of fluorescence and photochemistry: how, if at all, can different models of antenna organization be distinguished experimentally? Biochim Biophys Acta 1999; **1409**: 125–142. doi:10.1016/S0005-2728(98)00149-2

[3] Mullineaux CW, Pascal AA, Horton P, Holzwarth AR. Excitation-energy quenching in aggregates of the LHC II chlorophyll–protein complex: a time-resolved fluorescence study. Biochim Biophys Acta 1993; **1141**: 23–28. doi:10.1016/0005-2728(93)90184-H

[4] Bowers PG, Porter G. Quantum yields of triplet formation in solutions of chlorophyll. Proc R Soc A Math Phys Eng Sci 1967; **296**: 435–441. doi:10.1098/rspa.1967.0036

[5] Kok B, Forbush B, McGloin M. Cooperation of charges in photosynthetic O_2 evolution I. A linear four step mechanism. Photochem Photobiol 1970; **11**: 457–475. doi:10.1111/j.1751-1097.1970.tb06017.x

[6] Horton P. Are grana necessary for regulation of light harvesting? Au J Plant Physiol 1999; **26:** 659–669. doi:10.1071/PP99095

[7] Dall'Osto L, Lico C, Alric J, Giuliano G, Havaux M, Bassi R. Lutein is needed for efficient chlorophyll triplet quenching in the major LHCII antenna complex of higher plants and

effective photoprotection in vivo under strong light. BMC Plant Biol 2006; **6**: 32. doi:10.1186/1471-2229-6-32

[8] Foyer CH, Neukermans J, Queval G, Noctor G, Harbinson J. Photosynthetic control of electron transport and the regulation of gene expression. J Exp Bot 2012; **63** (4): 1637–1661. doi:10.1093/jxb/ers013

[9] Foyer CH, Noctor G. Redox sensing and signalling associated with reactive oxygen in chloroplasts, peroxisomes and mitochondria. Physiol Plant 2003; **119**: 355–364. doi:10.1034/j.1399-3054.2003.00223.x

[10] Foyer CH, Noctor G. Redox regulation in photosynthetic organisms: signalling, acclimation, and practical implications. Antioxid Redox Signal 2009; **11**: 861–905. doi:10.1089/ars.2008.2177

[11] Müller P, Li X-P, Niyogi KK. Non-photochemical quenching. A response to excess light energy. Plant Physiol 2001; **125**: 1558–1566. doi:10.1104/pp.125.4.1558

[12] Ruban AV, Berera R, Ilioaia C, van Stokkum IHM, Kennis JTM, Pascal AA, van Amerongen H, Robert B, Horton P, van Grondelle R. Identification of a mechanism of photoprotective energy dissipation in higher plants. Nature 2007; **450**: 575–578. doi:10.1038/nature06262

[13] Demmig-Adams B, Adams III WW. Photoprotection and other responses of plants to high light stress. Annu Rev Plant Physiol Plant Mol Biol 1992; **43**: 599–626. doi:10.1146/annurev.pp.43.060192.003123

[14] Horton P, Ruban A. Molecular design of the photosystem II light-harvesting antenna: photosynthesis and photoprotection. J Exp Bot 2004; **56** (411): 365–373. doi:10.1093/jxb/eri023

[15] Brunetti C, Guidi L, Sebastiani F, Tattini M. Isoprenoids and phenylpropanoids are key components of the antioxidant defense system of plants facing severe excess light stress. Environ Exp Bot 2015; **119**: 54–62. doi:10.1016/j.envexpbot.2015.04.007

[16] Liu Z, Yan H, Wang K, Kuang T, Zhang J, Gui L, An X, Chang W. Crystal structure of spinach major light-harvesting complex at 2.72 Å resolution. Nature 2004; **428**: 287–292. doi:10.1038/nature02373

[17] Bassi R, Pineau B, Dainese P, Marquardt J. Carotenoid-binding proteins of photosystem II. Eur J Biochem 1993; **212**: 297–303. doi:10.1111/j.1432-1033.1993.tb17662.x

[18] Wobbe L, Bassi R, Kruse O. Multi-level light capture control in plants and green algae. Trends Plant Sci 2016; **21**: 55–68. doi:10.1016/j.tplants.2015.10.004

[19] Horton P, Wentworth M, Ruban A. Control of the light harvesting function of chloroplast membranes: the LHCII-aggregation model for non-photochemical quenching. FEBS 2005; **579**: 4201–4206. doi:10.1016/j.febslet.2005.07.003

[20] Jahns P, Holzwarth AR. The role of the xanthophyll cycle and of lutein in photoprotection of photosystem II. Biochim Biophys Acta 2012; **1817**: 182–193. doi:10.1016/j.bbabio.2011.04.012

[21] Havaux M, Niyogi KK. The violaxanthin cycle protects plants from photooxidative damage by more than one mechanism. Proc Natl Acad Sci U S A 1999; **96**: 8762–8767. doi:10.1073/pnas.96.15.8762

[22] Mozzo M, Dall'Osto L, Hienerwadel R, Bassi R, Croce R. Photoprotection in the antenna complexes of photosystem II: role of individual xanthophylls in chlorophyll triplet quenching. J Biol Chem 2000; **283**: 6184–6192. doi:10.1074/jbc.M708961200

[23] Krüger TPJ, Ilioaia C, Johnson MP, Belgio E, Horton P, Ruban AV, van Grondelle R. The specificity of controlled protein disorder in the photoprotection of plants. Biophys J 2013; **105**: 1018–1026

[24] Stahl W, Sies H. Antioxidant activity of carotenoids. Mol Aspects Med 2003; **24**: 345–351. doi:10.1016/S0098-2997(03)00030-X

[25] Cazzaniga S, Li Z, Niyogi KK, Bassi R, Dall'Osto L. The Arabidopsis szl1 mutant reveals a critical role of β-carotene in photosystem I photoprotection. Plant Physiol 2012; **159:** 1745–1758. doi:10.1104/pp.112.201137

[26] Havaux M. Carotenoid oxidation products as stress signals in plants. Plant J 2014; **79**: 597–606. doi:10.1111/tpj.12386

[27] Dall'Osto L, Cazzaniga S, North H, Marion-Poll A, Bassi R. The Arabidopsis aba4-1 mutant reveals a specific function for neoxanthin in protection against photooxidative stress. Plant Cell 2007; **19**: 1048–1064. doi:10.1105/tpc.106.049114

[28] García-Plazaola JI, Becerril JM, Hernández A, Niinemets Ü, Kollist H. Acclimation of antioxidant pools to the light environment in a natural forest canopy. New Phytol 2004; **163**: 87–97. doi:10.1111/j.1469-8137.2004.01096.x

[29] Ariz I, Esteban R, García-Plazaola JI, Becerril JM, Aparicio-Tejo PM, Moran JF. High irradiance induces photoprotective mechanisms and a positive effect on NH_4^+ stress in *Pisum sativum* L. J Plant Physiol 2010; **167**: 1038–1045. doi:10.1016/j.jplph.2010.02.014

[30] Domonkos I, Kis K, Gombos Z, Ughy B. Carotenoids, versatile components of oxygenic photosynthesis. Prog Lipid Res 2013; **52**: 539–561. doi:10.1016/j.plipres.2013.07.001

[31] Aro E-M, Virgin I, Andersson B. Photoinhibition of photosystem II. Inactivation, protein damage and turnover. Biochim Biophys Acta 1993; **1143**: 113–134.doi:10.1016/0005-2728(93)90134-2

[32] Nishiyama Y, Murata N. Revised scheme for the mechanism of photoinhibition and its application to enhance the abiotic stress tolerance of the photosynthetic machinery. Appl Microbiol Biotechnol 2014; **98**: 8777–8796. doi:10.1007/s00253-014-6020-0

[33] Järvi S, Suorsa M, Aro E-M. Photosystem II repair in plant chloroplasts — regulation, assisting proteins and shared components with photosystem II biogenesis. Biochim Biophys Acta 2015; **1847**: 900–909. doi:10.1016/j.bbabio.2015.01.006

[34] Tyystjärvi E, Aro E-M. The rate constant of photoinhibition, measured in lincomycin-treated leaves, is directly proportional to light intensity. Proc Natl Acad Sci U S A 1996; **93**: 2213–2218. doi:10.1073/pnas.93.5.2213

[35] Melis A. Photosystem-II damage and repair cycle in chloroplasts: what modulates the rate of photodamage in vivo? Trends Plant Sci 1999; **4**: 130–135. doi:10.1016/S1360-1385(99)01387-4

[36] Carrillo N, Ceccarelli EA. Open questions in ferredoxin-NADP+ reductase catalytic mechanism. Eur J Biochem 2003; **270**: 1900–1915. doi:10.1046/j.1432-1033.2003.03566.x

[37] Guerrero MG, Vega JM, Losada M. The assimilatory nitrate-reducing system and its regulation. Annu Rev Plant Physiol 1981; **32**: 169–204. doi:10.1146/annurev.pp.32.060181.001125

[38] Nakayama M, Akashi T, Hase T. Plant sulfite reductase: molecular structure, catalytic function and interaction with ferredoxin. J Inorg Biochem 2000; **82**: 27–32. doi:10.1016/S0162-0134(00)00138-0

[39] Los DS, Murata N. Structure and expression of fatty acid desaturases. Biochim Biophys Acta 1998; **1394**: 3–15. doi:10.1016/S0005-2760(98)00091-5

[40] Sunil B, Talla SK, Aswani V, Raghavendra AS. Optimization of photosynthesis by multiple metabolic pathways involving interorganelle interactions: resource sharing and ROS maintenance as the bases. Photosynth Res 2013; **117**: 61–71. doi:10.1007/s11120-013-9889-z

[41] Arnon DI, Allen MB, Whatley FR. Photosynthesis by isolated chloroplasts. Nature 1954; **174**: 394–396. doi:10.1038/174394a0

[42] Munekage Y, Hashimoto M, Tomizawa CK-I, Endo T, Tasaka M, Shikanai T. Cyclic electron flow around photosystem I is essential for photosynthesis. Nature 2004; **429**: 579–582. doi:10.1038/nature02598

[43] Allen JF. Cyclic, pseudocyclic and noncyclic photophosphorylation: new links in the chain. Trends Plant Sci 2003; **8**: 15–19. doi:10.1016/S1360-1385(02)00006-7

[44] Krieger-Liszkay A, Feilke K. The dual role of the plastid terminal oxidase PTOX: between a protective and a pro-oxidant function. Front Plant Sci 2015; **6**: 1147. doi:10.3389/fpls.2015.01147

[45] Kargul J, Barber J. Photosynthetic acclimation: structural reorganisation of light harvesting antenna – role of redox-dependent phosphorylation of major and minor chlorophyll a/b binding proteins. FEBS J 2008; **275**: 1056–1068. doi:10.1111/j.1742-4658.2008.06262.x

[46] Pfannschmidt T. Chloroplast redox signals: how photosynthesis controls its own genes. Trends Plant Sci 2003; **8**: 33–41. doi:10.1016/S1360-1385(02)00005-5

[47] Fernadez AP, Strand A. Retrograde signaling and plant stress: plastid signals initiate cellular stress responses. Curr Opin Plant Biol 2008; **11**: 509–513. doi:10.1016/j.pbi.2008.06.002

[48] Rosso D, Bode R, Li W, Krol M, Saccon D, Wang S, Schillaci LA, Rodermel SR, Maxwell DP, Hüner NPA. Photosynthetic redox imbalance governs leaf sectoring in the *Arabidopsis thaliana* variegation mutants immutans, spotty, var1, and var2. Plant Cell 2009; **21**: 3473–3492. doi:10.1105/tpc.108.062752

Chapter 4

Effects on the Photosynthetic Activity of Algae after Exposure to Various Organic and Inorganic Pollutants: Review

Andreas S. Petsas and Maria C. Vagi

Additional information is available at the end of the chapter

http://dx.doi.org/10.5772/67991

Abstract

Algal studies remain necessary for risk assessment and their utility in ecotoxicology is the evaluation of lethal and sub-lethal toxic effects of potential toxicants on inhabitants of several ecosystems. Effects on algal photosynthetic apparatus caused by various chemical species have been extensively studied. The present chapter summarizes the published data concerning the toxicity of various organic and inorganic pollutants such as oils, pesticides, antifoulants and metals on photosynthesis of aquatic primary producers. Biochemical mode of action resulting in the disruption of photosynthesis depends on the chemical's nature and the characteristics of the exposed microorganism. Observed differences in response and sensitivity by different species to the same toxicant were attributed to several algal characteristics including photosynthetic capacity, pigment type, cellular lipid and protein content, and cell size. Single species bioassays either for one chemical alone or in mixture have been well reported and tolerance of both marine and freshwater water-column phytoplaktonic species has been examined. Adequate published information on multispecies tests (communities) in laboratory and field studies exists. However, risk assessment on photosynthesis of microbenthic periphyton is inadequate, though it is essential especially for hydrophobic organic molecules. Further studies are required to evaluate the adverse effects of metabolites on aquatic microalgae.

Keywords: aquatic toxicology, microorganisms, chlorophyll, photosynthesis, pollutants

1. Introduction

Aquatic ecosystems receive direct or indirect inputs of a wide diversity and a variety of chemical species among which polychlorinated biphenyls (PCBs), chlorinated dioxins, polycyclic aromatic hydrocarbons (PAHs), insecticides, herbicides, oils, metals and metalloids, inorganic nonmetallic elements, effluents, surfactants, synthetic detergents, and pharmaceuticals are included. Especially sediments (estuarine, river, and lake) accept the highest loads of all these aforementioned organic and inorganic molecules in both marine and freshwater aquatic environments. As a consequence, several compounds can play the role of toxic agents that inevitably expose inhabitants of these ecosystems which are vulnerable to pollution [1].

Fortunately, over the past few decades an enormous emphasis was placed on the section of aquatic toxicological research. Environmental protection agencies in a number of countries, particularly in Europe, North America, Japan, Southeast Asia, and Australia-New Zealand, in order to deal with wastewater discharges and in addition in their efforts to curb aquatic pollution, have recognized the great value of applying aquatic hazard assessment principles and procedures to effluents and their component chemicals and properties.

Phototrophic microorganisms such as micro- and macroalgae contribute significantly to primary productivity, nutrient cycling, and decomposition in the aquatic ecosystems; therefore, their importance in providing energy that sustains invertebrates and fish of those environmental compartments is very crucial. Microalgal communities form an essential functional group in aquatic habitats not only as key primary producers (important food source for feeders) but as regulators of oxygen levels; even at the water sediment interface, oxygen (O_2) production is highly dependent on the photosynthesis of microphytobenthos. Thus, the effects of toxic substances on algae are important not only for those microorganisms themselves but have subsequent impacts on higher trophic levels of the food chain. Since photosynthesis forms the fundamental basis of the food webs, even sub-lethal effects on primary producers could impact the energy transfer throughout the food chain [2].

As a result, toxicity tests have been developed that assess the effects of toxicants on photosynthetic activity of exposed species. The scientific published data demonstrate that the inhibition of photosynthetic activity is a common effect parameter monitored not only in numerous laboratory toxicity tests with cultured algae but also *in situ* with natural phytoplankton and periphyton communities [3].

The focus of this chapter is to provide a review of studies describing the toxicity of various organic and inorganic contaminants on the photosynthetic apparatus of aquatic microorganisms, such as algae. It describes the biochemical mode of action of each organic and inorganic pollutant concerning the disruption of photosynthesis, discusses the methods that have been employed for its analysis, compares the sensitivities of tested algal species to various toxicants, comments on the ecological relevance of the findings, and declines areas where future research is needed to be conducted.

2. Photosynthesis

Photosynthesis is an energy transformation process that converts light energy into chemical energy and is carried out by phototrophic organisms. Photosynthesis involves a series of biochemical and biophysical reactions occurring simultaneously in photosynthetic organisms (plants, algae, and cyanobacteria) that are always starting with the absorption of photons and ending with the incorporation of inorganic carbon into stable organic compounds called carbohydrates, such as sugars. The process of photosynthesis can be divided into two phases: the light reactions and the light independent or dark reactions. The light-dependent reactions of photosynthesis are mediated by four large protein complexes (also referred as supra-molecular complexes), embedded in the thylakoid membrane of the chloroplast: Photosystem I (PSI), Photosystem II (PSII), Cytochrome b_6/f Complex, and adenosine triphosphate (ATP) synthase [4]. In brief, light reactions involve the excitation of electrons of chlorophyll (chl) molecules within the PSII Complex to a higher energy state, which is the excited triple state ($^*chl^3$). This energy is harvested in the formation of several ATP molecules from ADP and inorganic phosphorus. In the PSI Complex, a similar excitation of electrons occurs, with the energy harvested to form reduced nicotinamide adenine dinucleotide phosphate (NADPH) from $NADP^+$. The electron transfer processes involved in the light-dependent reactions of photosynthesis are depicted in **Figure 1**, which is also known as Z-scheme of photosynthesis.

Algae during the light-dependent reactions of photosynthesis that take place in chloroplasts use pigment chl to absorb light, split the molecule of water, and therefore produce oxygen gas, and energy storage compounds of NADPH and ATP. Despite the fact that algae constitute a large, diverse, and polyphyletic group of organisms that exhibit enormous variations in morphology and physiology, the most important common biochemical attribute that unites photosynthetic algal species is their ability to perform photosynthesis.

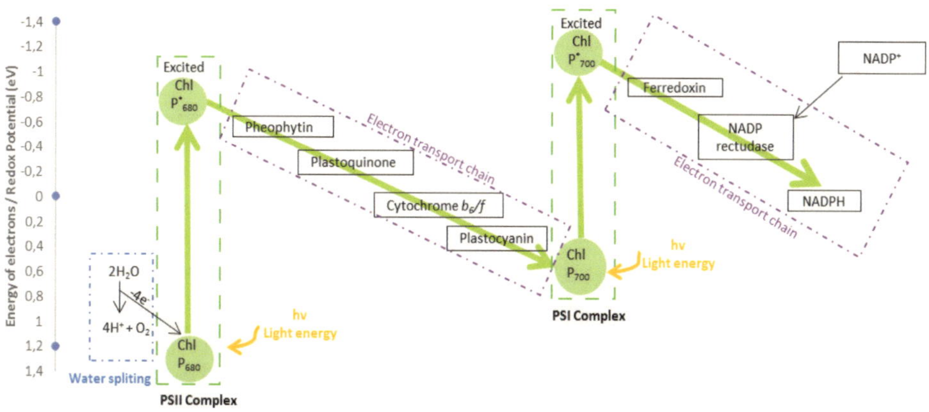

Figure 1. The Z-scheme of electron transfer processes involved in the light-dependent reactions of photosynthesis.

3. Methodologies of algal photosynthesis inhibition tests

Historically since the early 1900s, a variety of toxicity tests using algal species as exposed organisms have been performed for the evaluation of phytotoxic effects of several types of potential toxicants on aquatic inhabitants (including commercial chemicals, industrial and municipal effluents, and hazardous wastes). In the early 1970s and after taking into account the enormous ecological importance of bioassays, a number of regulatory and standard development agencies such as the Organization for Economic Cooperation and Development (OECD), International Standards Organization (ISO), European Economic Community (EEC), American Public Health Association (APHA), American Society for Testing and Materials (ASTM), and US Environmental Protection Agency (USEPA) developed and standardized phytotoxicity test methods. Current test methods are designed under the assumption that effects can be studied by three general approaches: (I) in a controlled laboratory experiment with limited number of variables, (II) in an experimental model ecosystem (indoor or outdoor simulator), and finally (III) in a natural ecosystem (*in situ*) [5].

Cause and effect relationships of specific chemicals to different types of target species are easily studied by the conduction of single-species laboratory-controlled experiments. The various methodologies of single-species bioassays differ slightly in design, but basically they utilize a uni-algal population of an available, easily cultivated, and sensitive algal test species (based on these criteria several microalgae have been recommended as standard test species, such as *Selenastrum capricornutum*), which is exposed during its log-growth phase to a range of concentrations of the toxicant [6].

The main disadvantage and limitation of single-species bioassays is the fact that they focus on assessing the effects of toxicants on single species and are performed under controlled laboratory conditions which are considerably different from the conditions of a realistic environment. In natural aquatic ecosystems, many complex species interactions and environmental influences and changes that cannot be simulated in laboratory studies continually occur. Other types of laboratory-conducted toxicological studies and beyond the level of single-species test are the multispecies tests and the small ecosystem tests, which are also called laboratory microcosms, and involve small-scale enclosures that contain natural samples (water, sediment, and algae) providing a simple simulation of natural systems. Phytoplankton and periphyton are the flora utilized in most multispecies toxicity tests [7].

Natural field studies or natural aquatic ecosystems tests (pond, stream, lake, or estuary) are defined as those in which both the test system and exposure to the stressor are naturally derived [8]. Field tests are very important and reliable for evaluating and understanding the biological and ecological effects of chemicals under real environmental conditions. Outdoor microcosms or mesocosms are simulated field studies that are composed of either an isolated subsection of the natural aquatic reservoir or a man-made physical model of an aquatic ecosystem, whereas the test systems are manually treated with the test chemical at predetermined test concentrations [8]. In general, the utilization of microcosms and mesocosms for assessing the effects of toxicants can reduce the possibility of an inaccurate estimation of the adverse effects of pollutants on aquatic species belonging to different ecological categories [9].

Photosynthetic activity is considered as a significant effect parameter of a variety of toxicants on algae (physiological and morphological effects). The primary advantage of photosynthesis tests is their short duration, which is usually 2–4 h, but exposure times have also ranged from 30 min to 24 h [7]. Therefore, the inhibitory and stimulatory effects of many organic and inorganic compounds on algal photosynthesis have been determined in laboratory and field studies. According to an extended published literature, several algal biochemical parameters linked to photosynthesis process such as ATP formation, CO_2 fixation, O_2 evolution, carbon uptake (^{14}C), and chlorophyll content have been adopted as traditional and classical indicators for the evaluation of environmental stresses caused by many classes of various contaminants on photosynthetic algal species [10].

A great progress in the area of algal photosynthesis research has been made during the last decades. Based on the fact that a proportion of the absorbed light energy in PSII photochemistry cannot be used to drive electron transport and is dissipated via non-radiative energy as heat or chlorophyll fluorescence emission associated with the PSII complex [2, 11–15], information about changes in the efficiency of photosynthesis can be acquired by measuring the yield of Chl-α-fluorescence [2, 16]. Chl-α-fluorescence is a physical signal defined as the radiative energy evolved from de-exciting Chl-α-molecules ($\lambda = 690$ nm for PSII, $\lambda = 740$ nm for PSI) [17] that has been used as a rapid, non-intrusive, and highly sensitive bioindicator of algal stress in response to different chemicals in recent years [2, 18, 19]. Apart from their utility in determining the physiological status of photosynthesizers in the natural environment, Chl-α-fluorescence-based methods are applied in ecophysiological and toxicological studies [2]. Among the various fluorescence techniques, pulse amplitude modulation (PAM) fluorometry, introduced by Schreiber et al. [11], has been demonstrated as a rapid, non-invasive, reliable, economically feasible, time-saving, and accurate technique, well suited for investigating changes in photochemical efficiency of aquatic algae, that permits *in vivo* non-destructive determination of changes in the photosynthetic apparatus much earlier than the appearance of visible damage [19]. Several types of PAM are known including the Maxi Imaging-PAM, Diving PAM, and ToxY-PAM fluorometer [2]. Numerous articles provide the efficiency of several Chl-α-fluorescence parameters that have been employed in assessing the effects of toxicants or their combinations on microalgae and macroalgae (seaweeds). Detailed definitions of certain Chl-α-fluorescence parameters along with their photosynthetic importance are available in the literature [16, 20–22]. The most commonly used Chl-α-fluorescence key parameters that are becoming recognized as valid sub-lethal indicators of photosystem stress and have been used to examine the sub-lethal toxicity of toxicants toward a variety of microalgae are maximum quantum yield (F_v/F_m), effective PSII quantum yield ($\Phi_{PSII,}$ or Φ_m or $\Delta F/F_{m'}$), operational PSII quantum yield (Φ'_{PSII} or Φ'_m), proportion of open PSII (qP), non-photochemical quenching (NPQ), and electron transport rate (ETR) [2, 23–26]. Hence, new types of devices of dual-channel PAM Chl fluorometers have been developed, which are specialized in the detection of extremely small differences in photosynthetic activity in algae or thylakoids suspensions. In conjunction with standardized algae cultures or isolated thylakoids, they provide an ultrasensitive bioassay system occupied frequently for the detection of toxic substances in water samples [24, 27]. Furthermore, many studies have directly compared the sensitivity of Chl-α-fluorescence end points to traditional indicators of organic and inorganic chemical stress on algae; these surveys include herbicides [26], antifouling agents, organometallic compounds [28], and metals [29].

4. Oils, dispersants, and dispersed oils

Naturally occurring raw or unprocessed crude oil and petroleum products are both included in the term "petroleum." Petroleum is a mixture of hydrocarbons of various molecular weights (most of which are alkanes, cycloalkanes, and various aromatic hydrocarbons), other organic compounds containing nitrogen, oxygen, and sulfur, and trace amounts of metals such as iron, nickel, copper, and vanadium. Hence, crude oil is a highly toxic compound comprising a mixture of up to 10,000 different types of hydrocarbons, both aliphatic and aromatic, which produce great damage to aquatic ecosystems [30]. On the other hand, processed and refined petroleum products include a large number of fuels, lubricants, and petrochemicals, such as gasoline, kerosene, diesel, paraffin wax, and many others that can cause important environmental contamination if released in ecosystems.

Hydrocarbons in aquatic environments have biogenic, natural geologic, and anthropogenic origins such as oil spills (releases of crude oil from tankers, offshore platforms, drilling rings, as well as spills of refined petroleum products and their by-products, or spills of any oil refuge or waste oil) [31–33]. Adverse effects resulting from spilled oil can be a result of (I) dissolved materials, (II) physical effects due to contact with oil droplets, (III) enhanced uptake of petroleum hydrocarbons through oil/organism interactions, or (IV) a combination of these factors [34]. Besides all the above, the insoluble and mainly the soluble fractions of oil reduce light penetration into the water column affecting phytoplankton photosynthesis process [35].

The ecological effects of accidental oil spills have been the subject of relevant laboratory and field research. Since the decade of 1950s, it has been known that crude and refined oils are phytotoxic [32], whereas the scientific interest concerning the sub-lethal effects of oils and their components on enzyme systems, photosynthesis, respiration, and protein and nucleic acid synthesis of primary producers is steadily increasing nowadays. According to published scientific data, it is demonstrated that toxic effect concentrations for oils and algae vary greatly. As previously reported in a recent review paper, the toxic effect concentrations range is between 0.002 and 10,000 ppm for crude oils and between 0.09 and 50 ppm for refined oils [32].

Based on information presented in the same bibliographic review of Lewis et al. on toxicity of oils, dispersants (mixtures of emulsifiers and solvents that break an oil slick into smaller droplets of oil), and dispersed oils toward algae and aquatic plants, 22 species of freshwater and 63 species of saltwater algae have been exposed to more oils (21) and dispersants (27) than any other type of aquatic plant [32]. This numeric example shows that even though damage may occur from low-level continuous discharges to both freshwater and saltwater environments, however, the environmental effects of large oil spills to marine waters have received the most attention by the public and regulatory and scientific communities resulting in the imbalance of entries in toxicity databases. Some of the available literature data concerning the toxicity of several types of oil or individual hydrocarbons on the photosynthetic apparatus reported for various algae are presented in **Table 1**.

The effects of crude oils and oil components on algae have been widely studied [43, 47–55], and among the different employed response parameters the effects on photosynthetic activity were included [43, 56, 57]. For that purpose, several algal species have been exposed to crude oils, fuel oils, dispersants, and dispersed oils not only in uni-algal cultures grown under laboratory-controlled conditions but also *in situ* as well by short- and long-term studies using microcosms, or mesocosms and mostly in short-term laboratory experiments. Toxicology studies

Test compounds	Test species	Observed stress response	References
Crude oils: South Louisiana, Kuwait, Venezuela, and Alaskan Diesel fuel oils No. 2: Amer. Petrol. Institute, Baton Rouge, Baytown, Montana, New Jersey	*Agmenellum quadriplicatum* *Chlorella autotrophica* *Cylindrotheca* sp.	Fuel oil: lethal at 10 mL $(20\ mL)^{-1}$. Crude oils: not toxic at 30 mL $(20\ mL)^{-1}$. Photosynthesis of *Chlorella autotrophica* was only temporarily depressed by the crude oils at 30 mL $(20\ mL)^{-1}$. Four of the fuel oils inhibited photosynthesis, O_2 output decreasing to zero without recovery (exception: Montana fuel oil).	Batterton et al. (1978) [36]
Crude oils: Atkinson Point, Norman Wells, Pembina, and Venezuela Corexit (unnamed)	*Laminaria saccharina* *Phyllophora truncata*	*In situ* primary production was significantly inhibited by all types and concentrations of oil tested (at 10 ppm). Inhibition generally increased with increasing oil concentration. The crude oil-Corexit mixtures were more toxic than crude oil or Corexit alone.	Hsiao et al. (1978) [37]
Coal liquefaction, shale-oil and petroleum products	*Selenastrum capricornutum* *Microcystis aeruginosa*	Based on ^{14}C assimilation measurements, the coal-liquefaction products inhibited algal photosynthesis at water-soluble fractions concentrations two orders of magnitude lower than the petroleum products; shale-oil products were intermediate in toxicity.	Giddings and Washington (1981) [38]
Crude oil: Tunisian	*Skeletonema costatum*	Toxicity is related to nutrient limitation conditions. 100 mg L^{-1} lethal in P and N limited media, and less severe in the Si-limited media. Chl-α and carbon uptake more sensitive parameters for assessing hydrocarbon toxicity than cell counting.	Karydis (1981) [39]
Crude oil: Ekofish	*Skeletonema costatum* *Phaeodactylum tricornutum* *Chaetoceros ceratosporum*	*S. costatum* and growth rate most sensitive than chlorophyll content per cell and the ratio of in vivo fluorescence to chlorophyll content.	Ostgaard et al. (1984) [40]
BP light diesel BP 1100X BP 1100WD Shell Oil Herder	*Chlorella salina*	Stimulatory effects on photosynthesis by low levels of BP light diesel (0.05%) and the oil dispersant BP 1100X (0.005%), either alone or in mixture. Inhibition of Chl-α content at higher levels of BP light diesel, BP 1100X and at all the tested concentrations of oil dispersants BP 1100WD and Shell Oil Herder.	Chan and Chiu (1985) [41]

Test compounds	Test species	Observed stress response	References
Crude oils: Ekofisk and Stratjford	*Skeletonema costatum* *Thalassiosira pseudonana* *Phaeodactylum tricornutum*	Reduced photosynthetic capacity. Highest sensitivity: *S. costatum.* Similar results by lab batch and *in situ* dialysis culture.	Hegseth and Ostgaard (1985) [42]
Crude oil: Norman Wells Corexit 9550	St. Laurence Estuary phytoplankton (*in situ* dosing)	Chl-α reduced at oil exposure concentration of 1–2 mg L^{-1}; No observed affection in marine community composition.	Siron et al. (1993) [43]
Diesel fuel oil No. 2: American Petroleum Institute	*Selenastrum capricornutum*	In terms of Chl-α content: 3d EC_{50} = 0.015 g L^{-1}; 5d EC_{50} = 0.014 g L^{-1}; 7d EC_{50} = 0.0156 g L^{-1}.	El-Dib et al. (1997) [44]
Chrysene (water soluble PAH)	Microcosms	Photosynthetic activity and chlorophyll-α concentration decreased after 24–72 h.	González et al. (2009) [35]
Oil samples from the tanker *Prestige* spill	*Dunaliella tertiolecta*	Significant inhibition of photosynthesis (based on F_v/F_m, ETR_{max}, and photosynthetic efficiency α-values) after only 1 h of oil exposure with clear concentration dependency. After 3 d, photosynthesis remained inhibited although cell survival was only slightly effected.	Carrera-Martinez et al. (2010) [30]
Eight groups of crude oil	Marine phytoplankton community	High concentrations of oil (≥2.28 mg L^{-1}) of decreased Chl-α content.	Huang et al. (2010) [45]
Accidental oil spill in Mumbai Harbor	Natural periphyton	*In situ*: significant decrease in phytoplankton population, inhibition of photosynthesis associated with degradation of pigments (increase in phaeophytin).	Jaiswar et al. (2013) [46]

Note: *Pseudokirchneriella subcapitata*, known as *Selenastrum capricornutum*.

Table 1. Examples of oils and hydrocarbons toxicity on the photosynthetic apparatus reported for various algae. Reports in chronological order.

conducted with photosynthetic aquatic communities usually indicate a shift of species composition and abundance after an oil spill due to the replacement of sensitive species by resistant ones (observations of short-term studies) [58]. Long-term studies in most cases reported cascades of late, indirect impacts on coastal communities due to chronic exposures to environment-sequestered petroleum products that delayed ecosystem recovery for years after an oil spill [59, 60]. Results of phytoplankton community studies are quite variable depending on characteristics of the oil, characteristics of the exposed algal species, influence of dispersants, type of ecosystem affected, dynamics of water masses, and numerous other variables [60, 61]. Therefore, the ecological impact following an oil spill depends on the volume spilled, oil type, geographical location of the spill, the characteristics of the receiving water, and its biota (e.g., sensitivity of organisms), and duration of contact with oil [62].

Short-term laboratory experiments, using laboratory-tolerant taxa and model experimental designs, have also been performed in order to evaluate more specifically the effects of different petroleum products on algal photosynthesis. Toxicity data obtained from laboratory assays indicate that toxic effects depend on the phytoplanktonic species, the group of oils involved, and the physical characteristics of the water, such as concentrations of dissolved organic compounds, temperature, salinity currents, redox potential, and nutrient loading [60].

In general, responses of microscopic photosynthesizers to oil are diverse [63]. In some case studies, growth rate has been shown as a more sensitive end point parameter than photosynthetic activity [40], whereas in others Chl-α-content and carbon uptake were more sensitive parameters for assessing hydrocarbon toxicity than cell counting [39]. In our knowledge, the dominant effect observed on photosynthetic activity after exposure to petroleum hydrocarbons is inhibition, while stimulation effects at low exposure levels of the toxicants have been also reported [37, 41, 64].

These findings are in accordance with the observations that microalgae have the capability to grow in the crude oil-contaminated environments, such as in the case of the rapid adaptation of mesophile species to crude oil of the Arroyo Minero River (Argentina) [30]. Hence, microalgae are able to survive in adverse environments as a result of physiological acclimation due to the modification of gene expression [30]. However, when values of environmental stress exceed physiological limits, survival depends exclusively on adaptive evolution, which is supported by the occurrence of mutations that confer resistance [30].

5. Pesticides

Pesticides are phytotoxins that are widely used all over the world in agriculture to kill unwanted vegetation. Pesticides are defined as substances or mixtures of substances intended for controlling, preventing, destroying, repelling, or attracting any biological organism deemed to be a pest. Insecticides, herbicides, defoliants, desiccants, fungicides, nematicides, avicides, and rodenticides are some of the many categories of pesticides. Many members of these compounds are very selective and are applied against certain target species, whereas many others are completely non-selective and thus effective to almost every species of plants acting as wide-spectrum molecules.

Paradoxically, these substances do not always remain in agricultural soils where they are applied for crop protection and fruit tree treatment, but sometimes they find their way into aquatic systems through leaching, surface runoff, spray-drift, soil erosion, and volatilization. Estimates indicate that the average agricultural herbicide loss is around 1% of the applied volume [27, 65]. In addition, millions of pounds of active pesticide ingredients are applied in coastal watersheds each year and that way pesticides may affect marine inhabitants via spills, runoff, and drift [66]. As a consequence, aquatic reservoirs receive direct and indirect pesticide inputs, inevitably exposing microorganisms to pesticides.

Pesticides have been classified by scientists according to their mechanisms of action. Photosynthetic inhibitors include many chemical groups of herbicides that disrupt photosynthesis pathways by four basic mechanisms that are summarized in **Figure 2**.

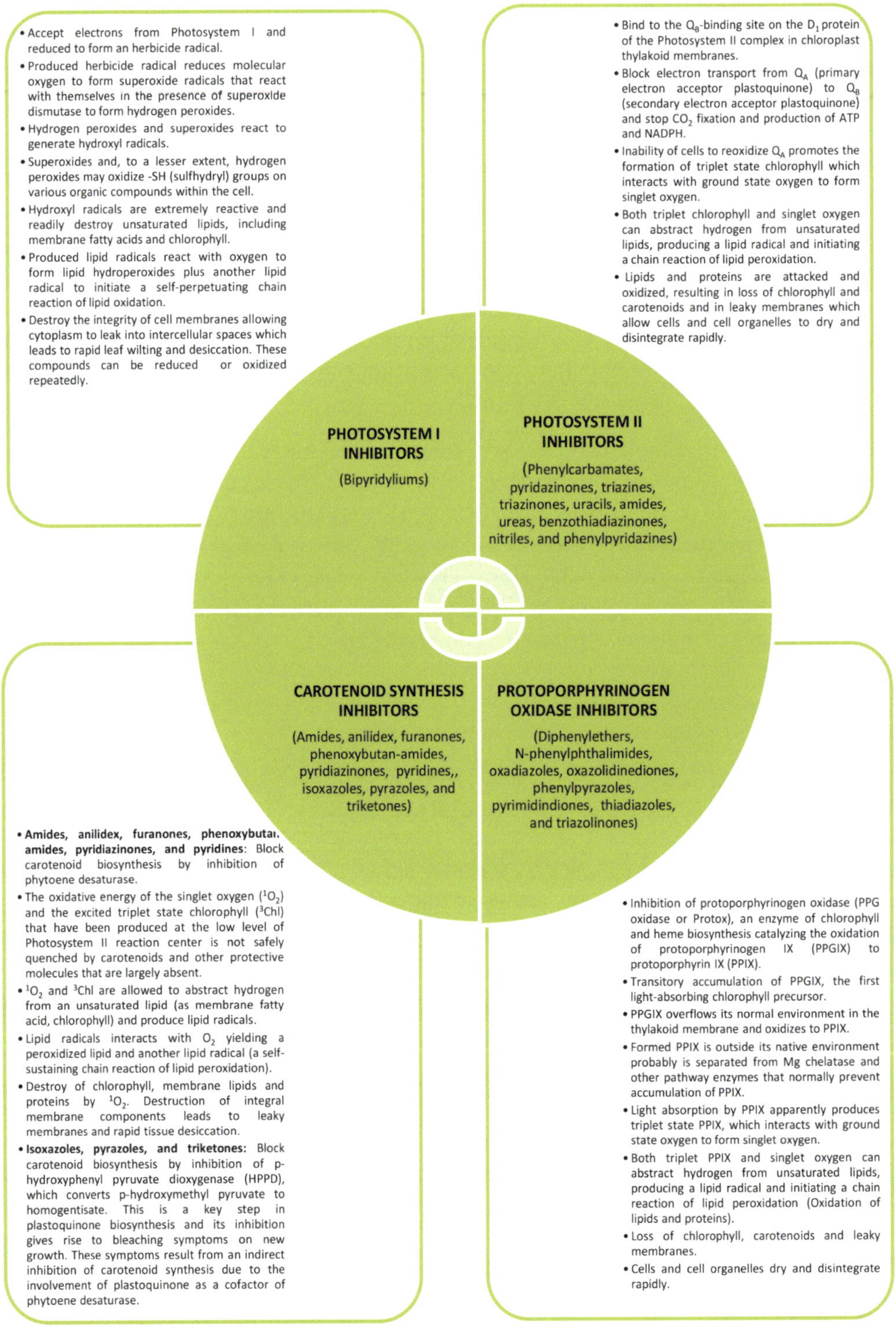

Figure 2. Photosynthetic inhibitors and their mechanism of action.

http://dx.doi.org/10.5772/67991

In this point, it must be mentioned that even though the majority of the pesticides is designed to and produced in the market with the assumption that they directly affect only one primary molecular site of action in the target organism; however, many of these compounds can cause a cascade of secondary and tertiary effects as well. For example, it has been found that most photosynthetic inhibitors also can affect plant respiration at higher doses [67]. Oxidative stress can also occur as a secondary effect of PSII inhibitors [68].

Furthermore, many non-photosynthetic inhibitors have been found to have an effect on photosynthetic process of various algal species. The herbicide flazasulfuron, a member of the chemical group of sulfonylureas, which are known to cause inhibition of amino acid synthesis, belongs to that case; bioassays conducted with the freshwater algae *Scenedesmus obliquus* revealed reduction in chlorophyll content at exposure concentration of 10 $\mu g\ L^{-1}$, while the increase of pigment content was reduced with the lowest tested level of exposure (0.1 $\mu g\ L^{-1}$) [69]. Moreover, studies of pesticide effects on algae showed that some pesticides can inhibit photosynthesis process with two independent mechanisms. For example, it has been reported that fluometuron, a substituted phenylurea compound, not only inhibited the production of Chl pigment in the unicellular algae *Chlorella pyrenoidosa* and *Euglena gracilis* but also blocked the biosynthesis of carotene via a process known as bleaching [70].

A broad base of toxicity data involving ecotoxicology of several classes of herbicides toward non-target microorganisms is available. Numerous reports have elaborated the impacts of various herbicides to algal photosynthetic activity. However, due to limited extent only few of them are selected to be presented herein this chapter. Therefore, only some of the available data in the literature are summarized in **Table 2** so as to depict the wide range among exposed algal species and among the employed photosynthesis parameters.

Algal species vary considerably in sensitivity to herbicides stress, and several factors may contribute to species-specific sensitivity including pigment type and photosynthetic capacity, cellular lipid and protein content, and cell size [71]. For instance, tolerance to atrazine has been linked to cell size in microalgae [71], whereas increased atrazine sensitivity to cell biovolume was observed, with smaller species being more sensitive to the herbicide [72]. What is more, algal subcellular responses to herbicides have been found to be also species dependent. In general, chlorophytes are considered to be more sensitive than bacillariophytes when comparing herbicide toxicity across phyla [73]. It has been well established that environmental parameters (light exposure, nutrient concentrations, etc.) interfere in the responses of algal communities to pesticides [74, 75]. As reported in reference [74], diatoms were more sensitive to atrazine during light exposure, suggesting that in the context of light, the response of algae depends on the season of study and on the site where samples are taken [76]. Light history has previously been implicated in periphytic (attached) microalgae, with shade-adapted (generally diatom-dominated) communities less susceptible than sun-adapted (chlorophyte-dominated) communities [74].

Additionally, in some species, results of algal bioassays may vary significantly based on the end point selected. As reported in a published comparative study of four estuarine microalgal species, a planktonic chlorophyte (*Dunaliella tertiolecta*), a benthic chlorophyte (*Ankistrodesmus* sp.), a cryptophyte (*Storeatula major*), and a dinoflagellate (*Amphidinium*

Pesticide (Chemical class)	**Test species**	**Exposure conditions, observed stress response and findings**	**References**
Glyphosate (Organophosphate)	Periphytic algal communities from 6 small forest ponds	Short-term carbon assimilation. Exposure range: 0.89–1800 mg L^{-1}. Photosynthetic activity decreased with increasing herbicide concentration in most ponds. Range of EC_{50} values: 8.9–89 mg L^{-1}.	Goldsborough et al. (1998) [77]
Flazasulfuron (Sulfonylurea)	*Scenedesmus obliquus*	24 or 48 h at 0.1–1000 μg L^{-1} (Chl-*α* and -*b*, carotenoids content): Reduction in chls content at 10 μg L^{-1}, while the increase of pigment content was reduced with the lowest tested level of exposure (0.1 μg L^{-1}). Among the three pigments studied Chl-*α* was the more sensitive biomarker.	Couderchet and Vernet (2003) [69]
Atrazine (Triazine)	*Dunaliella tertiolecta* *Ankistrodesmus* sp. *Storeatula major* *Amphidinium operculatum*	Nominal concentrations of atrazine tested: 0, 12.5, 25, 50, 100, and 200 μg L^{-1}. Atrazine significantly decreased cell density, productivity rate, biomass, and biovolume in all the algal populations tested at atrazine concentrations ≥12.5 μg L^{-1}. Based on photosynthetic carbon assimilation: *D. tertiolecta*: EC_{50} = 66.81 μg L^{-1}; *Ankistrodesmus* sp.: EC_{50} = 37.07 μg L^{-1}; *Storeatula major*: EC_{50} = 22.17 μg L^{-1}; *A. operculatum:* EC_{50} = 33.07 μg L^{-1}; Based on photosynthetic pigments content: *D. tertiolecta*: EC_{50} = 65.00 μg L^{-1}; *Ankistrodesmus* sp.: EC_{50} = 11.87 μg L^{-1}; *Storeatula major*: EC_{50} = 45.81 μg L^{-1}; *A. operculatum:* EC_{50} = 146.71 μg L^{-1}.	DeLorenzo et al. (2004) [71]
Cypermethrin (Pyrethroid)	*Scenedesmus obliquus*	96 h at 50–250 mg L^{-1} (Chl-*α* and -*b*, carotenoids content): Decreased contents of chls and carotenoids. Carotenoids production more sensitive than the ratio of Chl-*α*/Chl-*b*.	Li et al. (2005) [79]
Atrazine, simazine, hexazinone (Triazine) and diuron (Urea)	*Phaeodactylum tricornutum*	Based on PSII quantum yield: Atrazine: IC_{10} = 4.4 μL L^{-1}; Simazine: IC_{10} = 29.0 μL L^{-1}; Hexazinone: IC_{10} = 2.7 μL L^{-1}; Diuron: IC_{10} = 0.74 μL L^{-1}	Bengtson Nash et al. (2005) [27]
40 herbicides from 18 chemical classes and 9 modes of action	*Raphidocelis subcapitata*	EC_{50} with respect to the photosynthetic processes ranged from 0.0007 to 4.2286 mg L^{-1}. Descending order of the average acute toxicity was photosynthetic process>cell division>lipid synthesis, acetyl-coenzyme A carboxylase>acetolactate synthase> 5-enolpyruvyl-shikimate-3-phosphate-syntha-se, glutamine synthase, hormone synthesis>protoporphyrinogen oxidase.	Ma et al. (2006) [88]

Pesticide (Chemical class)	Test species	Exposure conditions, observed stress response and findings	References
Atrazine (Triazine), diuron, isoproturon (Ureas), paraquat dichloride (Bipyridinium)	*Selenastrum capricornutum*	Based on: Φ_m, Φ'_m, *NPQ* (1.5 h), F_{684}, F_{735} (30 min: Atrazine at concentrations 1.0–500 μg L^{-1} range or IC_{50} = 71.7–205.2 μL L^{-1}; Diuron at concentrations 0.2–100 μg L^{-1} range or IC_{50} = 7–12.3 μL L^{-1}; Isoproturon at concentrations 0.4–3.875 μg L^{-1} range or IC_{50} = 38.7–59.7μL L^{-1}; Paraquat dichloride at concentrations 2.0–1000 μg L^{-1} range of IC_{50} = 65.5–104.7 μL L^{-1}.	Fai et al. (2007) [10]
Diuron, hexazinone and atrazine (triazine/ triazinone)	*Navicula* sp. *Nephroselmis pyriformis*	The relationships between Φ_{PSII}, growth rate, and biomass increase were consistent ($r^2 \geq 0.90$) and linear (1:1). Order of toxicity (EC_{50} range) was diuron (16–33 nM) > hexazinone (25–110 nM) > atrazine (130–620 nm) for both algal species.	Magnusson et al. (2008) [26]
Mixture of diuron (phenylurea) and tebuconazole (triazole)	Natural periphyton in two series of two lotic outdoor mesocosms	The effects of pulsed acute exposures to pesticides on periphyton depended on whether the communities had previously been exposed to the same stressors or not.	Tlili et al. (2011) [89]

Note: *Pseudokirchneriella subcapitata*, known as *Selenastrum capricornutum*.

Table 2. Examples of pesticides toxicity on the photosynthetic apparatus reported for various algae. Reports in chronological order.

operculatum), which were exposed to atrazine, significant differences in sensitivity were observed depending on the test end point used. Chlorophyll-*α* was a significantly more sensitive test end point for *Ankistrodesmus* sp., biovolume was a significantly more sensitive test end point for *A. operculatum*, and phototrophic carbon assimilation was a significantly more sensitive test end point for *S. major* and *A. operculatum* [71]. In the same survey, it is suggested that species with greater Chl-*α* per cell are expected to be less sensitive to PSII inhibitors, because Chl-*α* is directly related to the amount of PSII in the cell, which is the primary biochemical target of such insecticides, and hence the more photosynthetic targets available, the more pesticide would be required to block it [71].

A dose-dependent inhibition of photosynthetic activity of algae has been reported in cases of single species [10, 27, 71] and as well as in periphytic algae exposures to a range of insecticides concentrations [77, 78].

According to the bibliographic data, available pigments content has often been used as a classic biomarker of exposure to pesticides in plants including algae and phytoplankton [69, 79, 80]. In other cases of published ecotoxicology studies evaluating the inhibition of photosynthesis by PSII inhibitors, Chl-*α*-fluorescence parameters were selected instead as test end points, emphasizing the precision and time-saving virtues of the technique [10, 24, 81]. For example, the inhibition of effective quantum yield (Φ_{PSII} or $\Delta F/F_{m'}$) has been used by many authors in order to examine the sub-lethal toxicity of herbicides toward a variety of microalgae, with some being sensitive to diuron at environmentally relevant concentrations [24, 25, 27]. Similar sensitivities were measured using ^{14}C uptake in benthic microalgae in temperate waters [82].

Taking into account the possible interactions between substances in combination, many mixture ecotoxicological experiments were performed using binary or ternary combinations of herbicides [83, 84]. Furthermore, a large body of literature data is available concerning the prediction of the joint effect of mixtures of pesticides based on their individual impacts and specific modes of action [85, 86]. Concentration addition (CA) and independent action (IA) model are the most commonly used models to predict mixture effects for similar- and dissimilar-acting compounds, respectively. Both theories assume enhanced effects with an increasing number of compounds and non-interaction between substances. Therefore, a deviation from the prediction indicates antagonism (weaker effects than predicted) or synergism (stronger effects) [87].

Pesticides are probably the most well-studied chemical group within ecotoxicological mixtures studies. This is not only due to the use of chemical mixtures in pesticide formulations and tank mixtures and the resulting co-occurrence in agricultural areas, but just as much because of the in-depth knowledge of their physiological mode of action [87]. These facts make them ideal candidates for testing mixture models based on the chemical mode of action and understanding the physiological mechanisms behind possible interactions [85, 90]. Mixture toxicity studies focused on single species [85, 86], natural communities in laboratory experiments [3, 82, 91], or outdoor microcosms and mesocosms [83, 92–94] data. Many reviews and critical analysis have shown that synergistic interactions within pesticide mixtures and realistic low-dose chemical mixtures in species are a rather rare phenomenon, constituting very low percentages of the tested mixture combinations and often occurs at high concentrations [87, 95–101]. According to the results of a comprehensive systematic review in which cocktail effects and synergistic interactions of chemicals in mixtures were predicted, synergy phenomena occurred only in 7% of the 194 binary pesticide mixtures included in the data compilation on frequency [101] (the database of Belden et al. [98] provided data on 207 pesticide mixtures of which 194 were binary and another 13 consisted of more than two pesticides). Results of the same study showed that PSII herbicides did not induce synergy in any of the 33 mixtures performed on algae in the pesticide database [101].

6. Antifouling biocides

Antifouling biocides are chemical substances that deter the microorganisms responsible for biofouling. Biofouling or biological fouling is the accumulation of microorganisms, plants, algae, or animals on wetted surfaces; hence, it can occur almost anywhere where water is present (marine vessels, swimming pools, drinking water and liquid lines for cooling electronics, medical devices and membranes, etc.). Biofouling takes place on surfaces after the formation of a biofilm that creates a surface onto which successively larger microorganisms can attach. Specifically designed antifouling materials and coatings/paints have the ability to remove or prevent biofouling by any number of organisms on such surfaces.

Antifouling biocides are introduced to antifouling paints in order to improve their efficacy against photosynthetic organisms [2]. The biocides often target the microorganisms which create the initial biofilm, typically bacteria. Other biocides are toxic to larger organisms in biofouling, such as the fungi and algae. Many different booster biocides have been currently added to antifouling paints including tributyltin (TBT), 2-methylthio-4-tetr-butylamino-6-cyclopropylamino-s-triazine (Irgarol 1051), 4,5-dichloro-2-n-octyl-4-isothiazolin-3-one (Sea-Nine 211), diuron, cuprous oxide, chlorothalonil, zinc pyrithione, dichlofluanid, 2,3,3,6-tetrachloro-4methylsulfonyl (TCMS), pyridine, 2-(thiocyanomethylthio) benzothiazole (TCMTB), and zineb [102].

One of the most commonly used biocides, and anti-fouling agents, is TBT. It is toxic to both microorganisms and larger aquatic organisms [103]. The mechanism of action of the TBT in algae is based on its interference with energy metabolism in chloroplasts and mitochondria, but it is also shown that TBT interacts with proteins and membranes and binds to or interacts with any protein containing free sulfhydryl groups [3, 104]. Bioassays conducted with the marine algae *Tetraselmis suecica* revealed that in chronic exposure to TBT, at higher concentrations (0.5–1 $\mu g\ mL^{-1}$) growth rate, chlorophyll pigments, carbohydrate, and protein contents were reduced [105]. Different responses have been described among three species of marine microalga *T. tetrathele*, *Nannochloropsis oculata*, and *Dunaliella* sp., which were exposed to three concentrations of TBT (0.1, 0.5, and 1$\mu g\ L^{-1}$). For *T. tetrathele*, exposure to TBT resulted in an increase of chlorophyll contents, even up to 210 and 225% at highest concentration of TBT (1$\mu g\ L^{-1}$) for chlorophyll α and b, respectively. However, acquired results for the other two algal species, *N. oculata* and *Dunaliella* sp., showed that stimulation effects occurred only at the lowest concentration tested (0.1 $\mu g\ L^{-1}$), as chlorophyll contents decreased at higher exposure levels, whereas *N. oculata* was the most sensitive microalga [106]. Similar results had been published in a previous study by Sidharthan et al. in which photosynthetic pigment content of the marine eustigmatophyte *N. oculata* was significantly affected, especially at elevated TBT concentrations. The same authors found that Chl-α content decreased more than 50% at TBT concentrations above 0.50 nM level, whereas at high concentration of 4 nM both the pigments were completely leached. Comparatively, carotenoid content was less inhibited by TBT toxicity (r = 0:917; $P < 0.05$) [107]. Reduction (60%) in the net photosynthetic activity of *Ruppia maritina* (seagrass) in TBT-spiked and impacted sediments was measured [108]. In a microcosm approach survey that was designed to study the combined effects of TBT from antifouling paints and ultraviolet-B radiation (UVBR: 280–320 nm), on a natural planktonic assemblage (<150 μm) isolated from the St. Lawrence Estuary (eastern Canada), it was demonstrated that phytoplankton cells were affected in their physiological functions, such as their photosynthetic efficiency. According to the obtained experimental data, the reduction in the maximum quantum yield (F_v/F_m) values were due to damage of PSII reaction centers and inhibition of ATP synthesis. Moreover, results clearly showed that the combination of TBT and UVBR stresses has synergistic effects affecting the first trophic levels of the marine food web [28]. Finally, the inhibition of photosynthesis of periphyton community has been observed after exposure to TBT ($EC_{50} = 0.02\ mg\ L^{-1}$) [3].

Irgarol 1051 is a triazine herbicide that has been described as an inhibitor of algal photosynthesis. More specifically, it belongs in PSII inhibitors, as it results in oxidative stress, including photooxidation of chlorophyll [109], and inhibition of the photosynthetic electron transport in chloroplasts by binding to the D_1 protein [110]. Irgarol 1051 was introduced after the restrictions on using TBT in antifouling paints (as a replacement) [111] and has found its application as an algicide in antifouling paints for boats and vessels. Irgarol is the most hydrophobic compound of the family of the triazines due to the presence of both tert-butyl group and the cyclopropyl group [102]. It is mainly used in combination with copper [3] and is the most frequently detected antifouling biocide worldwide [102]. Even though Irgarol 1051 is a relatively new compound, several papers have been published in the last years dealing with its ecotoxicological behavior toward non-target microorganisms. For example, in algal symbionts isolated from *M. mirabilis, D. strigosa,* and *F. fragum* 40–50% reduction of net ^{14}C incorporation has been demonstrated after their 6-h exposure to 10 mg L^{-1} of Irgarol 1051 [112]. Inhibition of the algal photosynthetic activity of several algal species including *D. tertiolecta, Synechococcus* sp., *E. huxleyi, Fucus vesiculosus, Enteromorpha intestinalis, Ulva intestinalis,* and seagrass *Z. marina* by Irgarol 1051 has been summarized [113]. In addition, the destruction of periphyton photosynthesis process after exposure to the same biocide has been demonstrated (EC_{50} = 0.82 nM) [114]. According to the available data, Irgarol 1051 has the potential to affect the F_v/F_m of phytoplankton even at very low (0.03 μg L^{-1}) environmentally relevant concentrations [115]. This conclusion is in accordance with the assumption that Irgarol 1051 concentration up to 0.23 mg L^{-1} negatively impacted the photosynthetic activity of the green alga *U. intestinalis* [116]. The effect of Irgarol on the values of several Chl-α-fluorescences parameters for numerous freshwater and marine algal species has been reported including the following data: according to F_v/F_m values: EC_{50} = 0.33 mg L^{-1} for *T. weissflogii*; EC_{50} = 0.60 mg L^{-1} for *E. huxleyi*; EC_{50} = 0.23 mg L^{-1} for *Tetraselmis* sp.; EC_{50} = 0.11 mg L^{-1} for *F. japonica* [117], reduction of F_v/F_m values in the presence of high concentrations for *Potamogeton pectinatus* [118]; whereas according to Φ_{PSII} or $\Delta F/F_{m'}$ values: 72 h EC_{50} = 0.327 mg L^{-1} for *T. weissflogii*; 72 h EC_{50} = 0.604 mg L^{-1} for *Emiliania huxleyi*; 72 h EC_{50} = 0.230 mg L^{-1} for *Tetraselmis* sp.; 72 h EC_{50} = 0.110 mg L^{-1} for *Fibrocapsa japonica* [119]; 72 h EC_{50} = 0.17 mg L^{-1} for *H. banksii* [120]; and 72 h EC_{50} = 2500 ng L^{-1} for *E. intestinalis* [121].

The other most commonly detected biocide in areas of high boating activity is diuron (phenylurea herbicide) [102]. The toxic effects of diuron on the photosynthetic apparatus of different algal species have been examined by many authors [10, 24–27, 89, 93, 115, 117] and among other ecotoxicological data the values of IC_{10} = 0.74 μL L^{-1} (based on PSII quantum yield) for *Phaeodactylum tricornutum* [27] and IC_{50} = 7 μL L^{-1} (based on $\Phi_{m'}$, 1.5 h) for *S. capricornutum* [10] are included. Natural periphyton studies have reported an induced increase in Chl-α content after long-term (29 days) exposure to low concentrations (1μg L^{-1}) of diuron [122]. This observation is in agreement with other previous studies of Tlili et al., who found that periphyton chronically exposed to 1 μg L^{-1} of diuron showed higher Chl-α pigments and carbon incorporation rates than control periphyton from day 21 to day 32 of their microcosm experiment [123]. That was confirmed in a more recent survey conducted in two series of two lotic outdoor mesocosms exposed to mixture of diuron and tebuconazole (triazole fungicide) which revealed induced tolerance to diuron, and therefore it was indicated that the effects of pulsed acute exposures to pesticides on periphyton depended on whether the communities had previously been exposed to the same stressors or not [89].

It has become well known that the antifouling biocide Sea-Nine 211 has an impact as an inhibitor of PSII electron transport [2, 113]. In addition, like other, more water-soluble representatives from the so-called Kathon group of biocides, Sea-Nine 211 quickly penetrates cell membranes and inhibits specific enzymes in the cell by reacting with intracellular thiols [3, 124]. Sea-Nine also seems to be able to affect more than one thiol group by generating a cascade of intracellular radicals [3]. Based on F_v/F_m measurements of natural phytoplankton communities, the toxicity of few biocides has been ranked as follows: Irgarol 1051 > zinc pyrithione>Sea-Nine 211>diuron. Thereby, it is suggested that Sea-Nine is more toxic than diuron, but less toxic than Irgarol [115]. In another survey, the toxicity of the antifoulants Sea-Nine, Irgarol, and TBT has been determined individually and in mixtures in two tests with microalgae and the effects on periphyton community photosynthesis and reproduction of the unicellular green algae *S. vacuolatus* have been investigated. The tested antifoulants have been found to be highly toxic in both tests. Observed mixture toxicities were compared with predictions derived from two concepts: independent action (IA) and concentration addition (CA), and IA failed to provide accurate predictions of the observed mixture toxicities. Mixture effects at high concentrations were slightly overestimated and effects at low concentrations were slightly underestimated [3].

Synergistic interactions have been foreseen not only between irgarol and diuron but between irgarol and chlorothalonil or 2-(thiocyanomethylthio)benzothiazole (TCMTB) as well. The synergies between irgarol and the two general fungicides, chlorothalonil and TCMTB, could be similar to the mechanism proposed for the PSII/metal interactions, as both fungicides create reactive oxygen species (ROS) and additionally chlorothalonil conjugates with glutathione, an important ROS scavenger [101].

7. Heavy metals and metalloids

In general, heavy metals are defined as metals with relatively high densities, atomic weights, or atomic numbers. On the basis of density, the term "heavy metal" is used for the elements that possess a density value greater than 4.5–5 g cm^{-3}, such as silver (Ag), arsenic (As), cadmium (Cd), copper (Cu), mercury (Hg), manganese (Mn), nickel (Ni), lead (Pb), and zinc (Zn), while metalloid is the definition of a chemical element that has properties intermediate between metals and non-metals, such as germanium(Ge), antimony (Sb), selenium (Se), tellurium (Te), polonium (Po), technetium (Tc), and astatine (At) [125].

Several metals are essential for living beings at very low concentrations, but at higher doses most of them are toxic for organisms belonging to different levels of the food chain [126]. Based on that criterion, metals are separated into the three following classes:

- **The essentials (class A)**: calcium (Ca), magnesium (Mg), Mn, potassium (K), sodium (Na), and strontium (Sr) (including macroelements which are metals that are required for algal growth, metabolism, and physiology (e.g., K and Mg) and microelements, which are metals that are required in trace amounts for certain biological processes and therefore must be obtained from the external environment).

- **The non-essentials (class B)**: Cd, Cu, Hg, and Ag.
- **The borderline class**: Zn, Pb, iron (Fe), chromium (Cr), cobalt (Co), Ni, As, vanadium (V), and tin (Sn) [127].

With regard to Ecotoxicology and Environmental Science, the term "heavy metals" is used to refer to metals that have caused environmental problems and includes chemical elements from the non-essentials and the borderline classes.

A steadily growing interest in the investigations on heavy metals is recorded and a large number of scientific surveys focused on the speciation of metals, their toxicity, accumulation, biomagnification, bioindication, migration, removal, phytoremediation, and biomonitoring have been conducted during the last decades. Cd, Hg, Zn, Cu, Ni, Cr, Pb, Co, V, titanium (Ti), Fe, Mn, Ag, and Sn are the metals that have been studied more extensively, whereas Hg, Cd, and Pb are some of the elements that have received the most scientific attention, possibly due to their highly toxic properties and their effectiveness on the environment and the living organisms [128].

Heavy metals can be naturally produced in aquatic system by the slow leaching from soil to water, usually at low levels [129]. Several other large natural inputs of heavy metals into water ecosystems are from the erosion or rocks, wind-blowing dusts, volcanic activity, and forest fires [128]. In addition, several anthropogenic activities such as energy production technologies, industrial effluents, and wastes (from coal mines, thermal power plants, metallurgy, plating, chemical plant, curry and paper-making industries, and other allied industries) alter the physicochemical characteristics of water bodies and elevate the heavy metals concentration according to the nature of effluent being discharged [130, 131]. Therefore, aquatic ecosystems receive inputs of different source containing a variety of metal ions (M^{x+}) that are directly or indirectly discharged into them.

Aquatic plants assimilate easily heavy metals, which are strongly phytotoxic and pose a threat to freshwater and marine life. Moreover, it has been well established that, depending on its bioaccumulation characteristics, a heavy metal can disperse through the various trophic levels of an ecosystem and its concentration levels are magnified [129]. Metals are not accessible to plants in their elemental forms (valence state of 0). On the contrary, they are available only in solution; hence, only metal ions play a role in biological systems [132]. The toxicity of metals and their compounds, however, largely depends on their bioavailability, that is, the mechanisms of uptake through cell membranes, intracellular distribution, and binding to cellular macromolecules [133]. In other words, the bioavailability of the metal, which depends on both biological factors and on the physicochemical properties of metallic forms (elements, their ions, and their compounds), is one of the key parameters in the assessment of the potential toxicity of metallic elements and their compounds toward organisms [125]. Metal availability is strongly dependent on environmental components, such as pH, redox and organic content, and soluble and bio-available metals. Hence, metals in the environment can be divided into two classes: (I) bio-available (soluble, non-sorbed, and mobile) and (II) non-bio-available (precipitated, complexed, sorbed, and no mobile).

Heavy metals enter algal cells by means of either active transport or endocytosis through chelating proteins and affect various physiological and biochemical processes of the algae. The mechanisms by which metals exert their toxicity on algae are very diverse and depend on the algal species, the nature and concentration of the metal, and the environmental conditions accompanying heavy metal stress [134]. Generally, their toxicity toward algal cells primarily results from (I) direct binding to the sulfhydryl groups (–SH) in functional proteins which disrupts their structure and function, and thus renders them inactive; (II) displacement of essential cations from specific binding sites that lead to a collapse of function; and (III) generation of reactive oxygen species, which consequently damages the macromolecules [126, 135].

At the sub-lethal level, heavy metals can interact with the vital process of photosynthesis. Interference of heavy metals with the photosynthesis of algae is a subject of intensive research that has been well documented. Almost all heavy metals are known to cause a negative impact on nearly all the components of the photosynthetic apparatus of primary producers [2, 132]. Direct effects of heavy metals on light and dark reactions and indirect effects resulting in the decrease of the photosynthetic pigment (including chlorophyll and carotenoid) content, as well as changes in stomata function, have been reported in the literature [132, 136]. Additionally, ions of heavy metal can damage the chloroplast membrane structure, disturb the light-harvesting and oxygen-evolving complexes, inhibit the photosystems and constituents of the photosynthetic electron transport chain, and also block the reductive pentose phosphate cycle [132, 137]. Moreover, toxic metals cause the inhibition of enzyme activities that are important in photosynthetic pathway. For example, it was found that Cd^{2+}, Zn^{2+}, and Hg^{2+} inhibited the NADP-oxidoreductase in *Euglena*, thereby significantly lowering the cell supply of NADPH [138], whereas Cu^{2+} was shown to inhibit plasma membrane H^+-ATPase activity in *Nitella flexis* [139]. Several enzymes involved in the Calvin cycle are also inhibited, especially Rubisco (bisphosphate carboxylase oxygenase) and PEPcarboxylase [132, 136]. Reaction of heavy metals with the enzyme-SH groups in proteins, substitution of essential ions, enhancement of photoinhibition and oxidative stress, impediment of plastocyanin function, change in lipid metabolisms, and disturbances in the uptake of essential microelements are other phenomena revealed due to heavy metal exposure [140, 141]. For instance, Cu^{2+} and Zn^{2+} substituted the Mg^{2+} in Chl molecules bound predominantly in the light-harvesting complex II of Chlorophyta, thereby impeding the PSII reaction centers, such as in the green alga *S. quadricauda* [141].

Finally, many heavy metals have been reported to influence the photosynthetic activity of algae through bleaching process. The observed bleaching effects have been connected with the tendency of toxic metals to generate ROS, such as singlet oxygen ($^1O_2$) and the hydroxyl radical (*OH), which can attack thylakoid lipids and initiate oxidation biochemical reactions that destroy membranes and damage structural pigment-protein complexes. For example, the toxicity of Cr^{6+} compounds has been traced to the reactive intermediates (formation of *OH radicals from H_2O_2 via a Fenton reaction) generated during the reduction of Cr by living cells [142]. As observed in the case of *Chlamydomonas reinhardtii* [134], this toxic metal tends to generate ROS, which can attack thylakoid lipids (mainly unsaturated fatty acids). This initiates peroxyl-radical chain reactions, destroying membranes and damaging indirectly structural pigment-protein complexes located in chloroplast membranes [2].

According to the numerous reported data on the photosynthesis inhibition by metals, three main experimental approaches can be distinguished: (I) results obtained from experiments with isolated chloroplasts or enzymes, to which heavy metals were supplied in the assay medium, (II) data acquired from experiments performed on excised leaves, exposed to a solution of the heavy metal, and (III) comparative laboratory experiments conducted on intact higher plants or algae, grown in a control medium and on a substrate enriched with heavy metals [140]. A summary of selected references on the toxicity of metals toward the photosynthetic apparatus for various microalgae is presented in **Table 3**.

Mercury is considered as the most toxic element among those having"no known physiological function" in algae. Based on results of ecotoxicological studies, Hg is recognized globally as an important pollutant and a serious threat to ecosystems. Hg and its compounds are persistent, bioaccumulative, and toxic. Inorganic Hg is the most common form of Hg released in the aquatic environment by industries [133]. Organic forms of Hg, such as methylmercury, revealed to have much stronger inhibitory effect than the inorganic mercury chloride on photosynthetic process [143]. Hg is able to alter the photosynthetic machinery including the chloroplastic PSI reaction center, subunit PSII, the oxygen-evolving protein, and the chloroplastic ATP synthase β-subunit [133, 144]. High levels of Hg in the form of Hg^{2+} have strong phytotoxic effects and when present in toxic concentrations can induce visible injuries and physiological disorders in plant cells triggering the production of ROS leading to cellular disruption [133].

Metallic form	Test species	Observed stress response	References
Cu^{2+}, Zn^{2+}	*Scenedesmus quadricauda* *Antithamnion plumula* *Ectocarpus siliculosus*	Under low irradiance heavy metal substitution of Mg in chl molecules bound predominantly in PSII of Chlorophyta; Under high irradiance the chls were inaccessible to substitution and the damage occurred in the PSII reaction center instead.	Kupper et al. (2002) [141]
Cu^{2+}, Ni^{2+}, Cd^{2+}, Zn^{2+}, Cr^{6+}	*Scenedesmus obliquus*	Inhibition of PSII photochemistry. Among the fluorescence parameters measured (after 12 h: F_o, F_v/F_m, qN, qP and after 1 h: F_m, $F_v/2$, and F_o/F_m) the highest sensitivity to all the five test metals had F_v/F_m.	Mallick and Mohn (2003) [29]
Co^{2+}	*Monoraphidium minutum* *Nitzschia perminuta*	Pigment content and photosynthetic O_2 evolution: increased at low levels and inhibited in high levels. Photosynthetic electron transport in *M. minutum* was more sensitive to Co^{2+} than in *N. perminuta*.	El-Sheekh et al. (2003) [156]

Metallic form	Test species	Observed stress response	References
CH_3Hg, Hg^{2+}	*Chlamydomonas reinhardtii*	$CH_3Hg \geq 1$ μM: Damaged the electron transfer chain at several sites; donor side of PSII, electron transfer from Q_A to Q_B, electron transfer between photosystems. Reduction of F_v/F_m, $\Delta F/F_m'$ and qN values. Hg^{2+} ($HgCl_2$) ≤ 5 μM did not affect F_v/F_m and $\Delta F/F_m'$ ratios.	Kukarskikh et al. (2003) [143]
Cd^{2+}, Cu^{2+}, Zn^{2}	*Dunaliella tertiolecta* *Promocentrum minimum* *Synechococcus* sp. *Thalassiosira weissflogii*	Comparable sensitivities of F_v/F_m and the cell-specific growth rate in quantifying the toxic effects of metals. *Synechococcus* sp. was the most sensitive species among the four algal species tested.	Miao et al. (2005) [145]
Ag^{1+}	*Chlamydomonas reinhardtii* *Pseudokirchneriella subcapitata*	Influence on proteins and enzymes for *C. reinhardtii* and on photosynthetic apparatus of *P. subcapitata.*	Hiriat-Baer et al. (2006) [157]
Cr^{6+}	*Chlamydomonas reinhardtii*	Complete pheophinitization of the chls and modification of the carotenoids.	Rodríguez et al. (2007) [134]
Cr^{6+}	*Eudorina unicocca* *Chlorella kessleri*	In *E. unicocca*: complete pheophinitization of the chls and modification of the carotenoids. In *C. kessleri*: no effect on the photosynthetic machinery even at higher levels of Cr^{6+}.	Juarez et al. (2008) [158]
Silver nano-particles (AgNP), Ag^{1+}	*Chlamydomonas reinhardtii*	Inhibition of photosynthesis by both AgNP and Ag^{+}. Based on total Ag concentration: Ag^{1+} ($AgNO_3$) displayed higher toxicity than AgNP. Based on Ag^{1+} concentration: AgNP displayed higher toxicity than Ag^{1+} ($AgNO_3$).	Navarro et al. (2008) [159]
Cu^{2+}, Cr^{6+}	*Euglena gracilis* (MAT and UTEX 753)	In the applied light conditions occurred, mainly damages to the PSII reaction center. Dark reactions were less sensitive.	Rocchetta et al. (2009) [150]
Cu^{2+}, Cr^{6+}, Zn^{2+}, Cd^{2+} Pb^{2+}	*Chlorella vulgaris*	Different effects on chl fluorescence for different metals: Cu and Cr had an inhibiting effect and Zn and Cd had a promoting effect.	Ou-Yang et al. (2012) [154]
Cd^{2+}	*Micrasterias denticulata*	Inhibition of PSII activity. Reduction of O_2 production. Structural damage of the chloroplast. Disturbance of Ca homeostasis by displacing Ca.	Andosch et al. (2012) [160]

Metallic form	Test species	Observed stress response	References
Cd^{2+}, Cu^{2+}, Zn^{2+}	*Planothidium lanceolatum* (Brébisson)	Significant effect on F_v/F_m at concentrations of $Cd^{2+} \geq 0.1$, $Zn^{2+} \geq 0.2$, and $Cu^{2+} \geq 0.4$ mg L^{-1}.	Sbihi et al. (2012) [155]
Cd^{2+}, Cr^{6+}, Cu^{2+}, Zn^{2+}	*Pseudokirchneriella subcapitata*	Modification of mitochondrial membrane. Reduction of photosynthetic activity.	Machado et al. (2015) [161]
Hg^{2+}	*Gracilaria salicornia* *Sargassum* sp. *Ulva reticulata*	Reduction of F_v/F_m and Chl-α content.	Bakar et al. (2015) [133]
Cu^{2+}, Pb^{2+}	*Gracilaria edulis* *Gracilaria manilaensis* *Gracilaria salicornia*	Reduction of the algal F_v/F_m in both metals. Cu^{2+} induced the synthesis of chl-a in *G. edulis* and *G. salicornia* but inhibited chl-α synthesis in *G. manilaensis*. Pb^{2+} induced the production of Chl-α in all tested algae.	Bakar et al. (2015) [162]
Pb^{2+}	*Anabaena* sp.	Reduction of pigment content (Chl-α and car) and photosynthetic efficiency (F_v/F_m) of PSII.	Deep et al. (2016) [163]

Note: *Pseudokirchneriella subcapitata*, known as *Selenastrum capricornutum*.

Table 3. Examples of metals toxicity on the photosynthetic apparatus reported for various algae. Reports in chronological order.

Copper is unquestionably an essential element in various metabolic processes of algae, such as amine oxidase and cytochrome c oxidase system, prosthetic group of the chloroplastic antioxidant enzyme Cu/Zn superoxide dismutase, and regulator of PSII-mediated electron transport. However, Cu is still considered as one of the most toxic heavy metal ions to algae and is a potent inhibitor of photosynthesis [2]. Many studies have examined ecotoxicological effects of Cu on photosynthetic activity of plants and phytoplankton [145]. From an evaluation of the literature, Cu can affect photosynthetic electron transport on the reducing side of PSI at the level of the ferredoxin [146], alter the PSII on the oxidizing side by inhibiting the electron transport at P680 (the primary donor of PSII) or by inactivating some PSII reaction centers [147]. Cu may also impair the PSII electron transport on its reducing side by affecting the rate of oxidoreduction [148]. The inhibitory effect of copper on the photosynthetic apparatus of several species of algae has been examined, including *E. gracilis* [149, 150], *S. quadricauda* [141], *S. obliquus* [151], *S. incrassatulus* [152], *C. pyrenoidosa* [153], *C. vulgaris* [154], *Planothidium lanceolatum* and *Isochrysis galbana* [155], *D. tertiolecta, Promocentrum minimum, Synechococcus* sp., and *Thalassiosira weissflogii* [145].

Cadmium is a heavy metal that occurs naturally in ores along with zinc, lead, and copper. Its compounds are used as stabilizers in PVC products, color pigment, several alloys, and in rechargeable nickel-cadmium batteries. Cd forms complexes with various organic particles and thereby triggers a wide range of reactions that collectively put the aquatic ecosystems to risk [2]. Due to its high toxicity at low concentration, Cd is considered as an important contaminant of natural waters [164]. Research regarding the adverse effects of Cd on microorganisms demonstrated that Cd^{2+}, via a variation of mechanisms, affected several biochemical algal processes. References include the displacement of $Zn^{2+}$and Ca^{2+} co-factors from undefined protein targets or directly binding amino acid residues, including cysteine, glutamate, aspartate, and histidine [165]; the inhibition of chlorophyll formation and the reduction of both chlorophyll content and Chl a/b ratio through disturbances in the electron transport chain in both PSI and PSII; and the reduction of Rubisco and enhancement of lipoxygenase activity [2, 145].

Chromium is a transition element that comprises the seventh most abundant metal in the earth's crust, whereas trivalent (Cr^{3+}) and hexavalent (Cr^{6+}) ions are its two most common and stable oxidation states in the environment. Whereas Cr^{3+} is considered a micronutrient, essential for the proper function of living organisms, Cr^{6+} instead can display numerous toxic effects on biological systems. Cr^{6+} is usually associated with oxygen to form chromate (CrO_4^{2-}) or dichromate ($Cr_2O_7^{2-}$) oxyanions that can easily go through cell membranes as an alternative substrate for the sulfate transport system and exhibit strong oxidative potential [166]. Therefore, Cr^{6+} is associated with several intracellular and ultra-structural modifications, among which the inhibition of photosynthesis is included. As observed in the cases of the algal species *Chlamydomonas* [134], *C. pyrenoidosa* [167], *Eudorina unicocca, C. Kessleri* [168], *E. gracilis* [150], *S. obliquus* [169], and *Monoraphidium convolutum* [170], Cr^{6+} caused an enhanced destruction of the reaction centers and a reduction in measured Chl-α-fluorescence parameters such as Φ_{PSII}, F_v/F_m, Φ'_{PSII}, *ETR*, and *qP* [2].

Zinc is an essential element for the activity of several enzymatic systems of organisms. Stimulatory effects on algal photosynthesis at low exposure concentrations of Zn^{2+} have been observed. For example, *C. vulgaris* after 96 h of exposure at treatment concentration of 5 μmol L^{-1} showed that the proportion of the maximum quantum yield of PSII promoted by Zn was approximately 10% [154]. However, when the external concentration of Zn^{2+} is beyond a limited value, it causes harmful effects; hence, its concentration in the cells must be controlled. Zn deficiency in *E. gracilis* has been shown to affect growth, morphology, cell cycle, and mitosis. These observations are best explained by a role for zinc in gene regulation, through zinc-dependent enzymes [149]. Significant effect on F_v/F_m ratio of *P. lanceolatum* (Brébisson) at a concentration level of 0.2 mg L^{-1} of Zn^{2+} was observed, while the sensitivity of the same algal species toward all tested heavy metals was diminishing in the order: $Cd^{2+} > Zn^{2} > Cu^{2+}$ [155].

The toxicity of ionic silver to a variety of aquatic organisms, such as algae, has been studied and shown to be significant, whereas from an evaluation of the literature, Ag+ displayed toxicity to aquatic photosynthetic microorganisms in the nanomolar (nM) concentration range [157, 159]. The toxicity of other forms of silver, such as silver nanoparticles (AgNP) ranged in size from 10 to 200 nm, has been examined as well and according to fluorometry values AgNPs were found to influence the photosynthesis of *C. reinhardtii* as well as ionic silver (Ag^+) [159].

At this point, it must be mentioned that due to the fact that aquatic ecosystems act as reservoirs of several mixtures of metals, it is essential to evaluate the combined or cumulative effect of metals or metal mixtures on photosynthesis. Therefore, toxicological studies dealing with heavy metal pollution in aquatic organisms must take into account the interactions among metals that may influence uptake, accumulation, and toxicity [2, 128]. For instance, it has been reported that interactions between Cu^{2+} and Mg^{2+} may have special significance regarding phytoplankton growth [2]. In another survey assessing the effect of Cu^{2+}, Cr^{6+}, and Ni^{2+} on growth, photosynthesis, and chlorophyll, a synthesis of *C. pyrenoidosa,* it was demonstrated that various bimetallic combinations of those metals interacted synergistically [171]. Combined effects of Cu^{2+} and Cd^{2+} on the growth and photosynthesis-related gene transcription of *C. vulgaris* have been also investigated [154].

In a more realistic approach, metals could also occur along with other contaminants in mixtures. In that respect, synergistic interactions have been predicted between pesticides that act as PSII inhibitors (and are included in the database of Belden et al. [98]) and the metals Cd, Cu, and Zn [101]. A proposed synergistic mechanism between metals and PSII inhibitors in autotrophs could be that metals might prevent the repair of not only damaged PSII complexes, which are constantly repaired during photosynthesis, but also the damage caused by the reactive oxygen species (ROS) created by the PSII inhibition and the metals themselves, by interacting with enzymes responsible for the repair [101].

Finally, metal bioassays must take into account the synthetic organometallic compounds or the ones formed under environmental conditions. These organometallic substances, especially of Hg, Pb, and Sn, might have completely different toxicological properties and can be more toxic to aquatic organisms because of their high bioaccumulation, as is the cases of methyl mercury compounds (methylation process is thought to be bacterially mediated) [128, 143] and tributyltin chloride [3, 105].

However, it must be underlined that several metal-tolerant algal strains, which have been adapted to environments contaminated with toxic metals (such as Cu and Cd), have been isolated and identified and a variety of tolerance mechanisms have been described [172]. Metallothioneins (MTs) consist one of the most important cellular defense mechanisms against metal stress that regulate the toxicity of various metals and trace elements. MT is a family of cysteine-rich and low-molecular-weight proteins localized to the membrane of the Golgi apparatus, which have the ability to bind several metals through the thiol clusters of their cysteine residues [173]. Some algal MTs are gene products, while others are secondary metabolites [172]. According to relevant studies, these molecules chelate toxic trace metals, for example, Cd, thereby reducing the concentration of cytotoxic, free-metal ions. Furthermore, some MTs are believed to be involved in zinc and copper homoeostasis [172]. The removal of heavy metals from polluted waters by the use of algae (e.g., *C. pyrenoidosa* and *Scenedesmus* sp.) is called phycoremediation and is an expanding technology with several advantages over physical remediation methods [174].

8. Conclusions and trends

One of the common and main goals of environmental science and ecotoxicology is the environmental sustainability that concerns the natural aquatic ecosystems and how they endure and remain diverse and productive. Taking into account that photosynthetic microorganisms are the main primary producers and consist of the basis of the food chains, a large number of toxicity tests have been conducted in order to assess the effects of a variety of environmental pollutants on algal photosynthetic activity. According to the available vast information, several bioassays have been performed with a great variety of standard test species of both freshwater and saltwater algae, though various "non-standard" algal species have been used on occasion. In our knowledge, in most cases freshwater microalgae were used more frequently in laboratory toxicity tests than any other types of aquatic plant, except in the case of oil spills where more data for marine algae are available. Moreover, literature data showed that the most commonly used microalgae in marine toxicity tests are green algae and diatoms. The observed differences in response and sensitivity by various microalgal and macroalgal species to the same toxicant can be several orders of magnitude for toxicants such as crude oils, oil products, pesticides, antifouling biocides, and metals. Evidenced heterogeneous sensitivity of different algal species to the same pollutant is attributed to several characteristics of the exposed alga such as photosynthetic capacity and pigment type, cellular lipid and protein content, and cell size.

Algae have been suggested and used as potential bioindicators of aquatic pollution [1, 175]. Damage of their photosynthetic apparatus is a very sensitive response to xenobiotics that could point to an important biomarker [79]. Carried out studies confirmed that inhibition of photosynthesis is one basic reflex of the toxic effects of several organic and inorganic pollutants on microalgae which in many cases is a more sensitive end point than inhibition of growth [39]. Therefore, we can conclude that measuring the photosynthetic activity is a good screening method for detecting a variety of possible stress situations [132].

Loadings of several anthropogenic pollutants are usually nearly and chronically synchronous with discharges, leading to marked changes in exposure levels of inhabitants of aquatic reservoirs. Depending on the nature, concentration, frequency, and duration of toxicants exposure, their impacts on biological communities can prove highly variable [89]. Until nowadays, many experimental studies of aquatic communities of microorganisms have been done using water-column phytoplanktonic species, but only a few have attempted to assess the effect of environmentally realistic pollution exposure scenarios on microbenthic periphyton [89, 122, 123, 176]. The distribution characteristics of chemical toxicants between water phase and sediment are of major importance in the evaluation of their fate and ecotoxicological effects into environmental compartments, especially for organic hydrophobic pollutants. Therefore, more vivid studies need to be performed in the future on the bioavailability of organic pollutants and the possible link between pollutant dynamics in the adsorbed phase (bottom sediment periphyton matrices) and their impacts on microbenthic photosynthetic algae.

Last but not least, there is still not much known about the possible toxic effects of transformation and degradation products of several synthetic organic compounds on aquatic microalgae. This lack of data makes the toxicity assessment of formed organic molecules metabolites essential, because these molecules may be more toxic than the parent ones; hence, further studies are required to evaluate the adverse effects of these produced chemical species on algal photosynthetic activity.

Author details

Andreas S. Petsas[1,2*] and Maria C. Vagi[2]

*Address all correspondence to: apetsas@env.aegean.gr

1 Department of Food Science and Nutrition, School of Environment, University of the Aegean, Myrina, Lemnos, Greece

2 Laboratory of Chemical Processes & Aquatic Toxicology, Department of Marine Sciences, School of Environment, University of the Aegean, University Hill, Mytilene, Lesvos, Greece

References

[1] Torres MA, Barros MP, Campos SCG, Pinto E, Rajamani S, Sayre RT, Colepicolo P. Biochemical biomarkers in algae and marine pollution: a review. Ecotoxicology and Environmental Safety. 2008;**71**:1–15. DOI: 10.1016/j.ecoenv.2008.05.009

[2] Kumar KS, Dahms HU, Lee JS, Kim HC, Lee WH, Shin KH. Algal photosynthetic response to toxic metals and herbicides assessed by chlorophyll α fluorescence. Ecotoxicology and Environmental Safety. 2014;**104**:51–71. DOI: 10.1016/j.ecoenv.2014.01.042

[3] Arrhenius A, Backhaus T, Grönvall F, Junghans M, Scholze M, Blanck H. Effects of three antifouling agents on algal communities and algal reproduction: mixture toxicity studies with TBT, Irgarol, and Sea-Nine. Archives of Environmental Contamination and Toxicology. 2006;**50(3)**:335–345. DOI: 10.1007/s00244-005-1057-9

[4] Dekker JP, Boekema EJ. Supramolecular organization of thylakoid membrane proteins in green plants. Biochimica et Biophysica Acta-Bioenergetics. 2005;**1706(1–2)**:12–39. DOI: 10.1016/j.bbabio.2004.09.009

[5] Rand GM, Wells PG, McCarthy LS. Introduction to aquatic toxicology. In: Rand GM, editor. Fundamentals of Aquatic Toxicology-Effects, Environmental Fate, and Risk Assessment. 2nd ed. Boca Raton: CRC Press; 1995. pp. 3–67. ISBN: 156032-091-5

[6] Lewis MA. Use of freshwater plants for phytotoxicity testing: a Review. Environmental Pollution. 1995;**87**:319–336. DOI: 10.1016/0269-7491(94)P4164-J

[7] Lewis MA. Algae and vascular plant tests. In: Rand GM, editor. Fundamentals of Aquatic Toxicology-Effects, Environmental Fate, and Risk Assessment. 2nd ed. Boca Raton: CRC Press; 1995. pp. 135–169. ISBN: 156032-091-5

[8] Graney RL, Giesy JP, Clark JR. Field studies. In: Rand GM, editor. Fundamentals of Aquatic Toxicology-Effects, Environmental Fate, and Risk Assessment. 2nd ed. Boca Raton: CRC Press; 1995. pp. 257–305. ISBN: 156032-091-5

[9] Velki M, Hackenberger BK, Lončarić Z, Hackenberger DK. Application of microcosmic system for assessment of insecticide effects on biomarker responses in ecologically different earthworm species. Ecotoxicology and Environmental Safety. 2014;**104**:110–119. DOI: 10.1016/j.ecoenv.2014.02.006

[10] Fai PB, Grant A, Reid B. Chlorophyll a fluorescence as a biomarker for rapid toxicity assessment. Environmental Toxicology and Chemistry. 2007;**26(7)**:1520–1531. DOI: 10.1897/06-394R1.1

[11] Schreiber U. Detection of rapid induction kinetics with a new type of high-frequency modulated chlorophyll fluorometer. Photosynthesis Research. 1986;**9:**259–270. DOI: 10.1007/978-94-009-4412-1_24

[12] Schreiber U, Schliwa U, Bilger W. Continous recording of photochemical and non-photochemical chlorophyll fluorescence quenching with a new type of modulation fluorometer. Photosynthesis Research. 1986;**10(1)**:51–62. DOI: 10.1007/BF0002418

[13] Genty B, Briantais JM, Baker NR. The relationship between the quantum yield of photosynthetic electron transport and quenching of chlorophyll fluorescence. Biochimica et Biophysica Acta-General Subjects. 1989; **990(1)**: 87–92. DOI: 10.1016/S0304-4165(89)80016-9

[14] Pullerits T, Sundström V. Photosynthetic light-harvesting pigment-protein complexes; towards understanding how and why. Accounts of Chemical Research. 1996;**29(8)**:381–389. DOI: 10.1021/ar950110o

[15] Popovic R., Davis DE. Applications of chlorophyll fluorescence in ecotoxicology: heavy metals, herbicides, and air pollutants. In: DeEll JR, Toivonen PMA editors. Practical Applications of Chlorophyll Fluorescence in Plant Biology. Berlin: Springer, 2003. pp. 151–184. ISBN: 978-1-4615-0415-3

[16] Maxwell K, Johnson GN. Chlorophyll fluorescence-A practical quide. Journal of Experimental Botany. 2000;**51(345)**:659–668. DOI: 10.1093/jexbot/51.345.659

[17] Roháćek K, Soukupová J, Barták M. Chlorophyll fluorescence: a wonderful tool to study plant physiology and plant stress. In: Schoefs B editor. Plant: Cell Compartments-Selected Topics. Kerala, India, Research Signpost. 2008. pp. 41–104. ISBN: 978-81-308-0104-9

[18] Muller R, Schreiber U, Escher BI, Quayle P, Bengston-Nash SM, Mueller JF. Rapid exposure assessment of PSII herbicides in surface water using a novel chlorophyll *a* fluorescence imaging assay. 2008;**401(1–3)**:51–59. DOI: 10.1016/j.scitotenv.2008.02.062

[19] Kumar KS, Han T. Physiological response of *Lemna* species to herbicides and its probable use in toxicity testing. Toxicology and Environment Health Science. 2010;**2(1)**:39–49. DOI: 10.1007/BF03216512

[20] Nedbal L, Soukupová J, Kaftan D, Whitmarsh J, Trtilek M. Kinetic imaging of chlorophyll fluorescence using modulated light. Photosynthesis Research. 2000;**66(1–2)**:3–12. DOI: 10.1023/A:1010729821876

[21] Roháćek K. Chlorophyll fluorescence parameters: the definitions, photosynthetic meaning, and mutual relationships. Photosynthetica. 2002;**40(1)**:13–29. DOI: 10.1023/A:1020125719386

[22] Elahifard E, Ghanbari A, Mohassel MHR, Zand E, Kakhki AM, Abbaspoor M. Measuring chlorophyll fluorescence parameters for rapid detection of ametryn resistant. Plant Knowledge Journal. 2013;**2(2)**:76–82. Available from: http://profdoc.um.ac.ir/paper-abstract-1034975.html [Accessed: 2017-01-10]

[23] Jones RJ, Kildea T, Hoegh-Guldberg O. PAM chlorophyll fluorometry: a new in situ technique for stress assessment in Scleractinian corals, used to examine the effects of cyanide from cyanide fishing. Marine Pollution Bulletin. 1999;38:864–874. DOI: 10.1016/S0025-326X(98)90160-6

[24] Schreiber U, Müller JF, Haugg A, Gademann R. New type of dual channel PAM chlorophyll fluorometer for highly sensitive water toxicity biotests. Photosynthesis Research. 2002;**74(3)**:317–330. DOI: 10.1023/A:1021276003145

[25] Escher BI, Quayle P, Muller R, Schreiber U, Mueller JF. Passive sampling of herbicides combined with effect analysis in algae using a novel high-throughput phytotoxicity assay (Maxi-Imaging-PAM). Journal of Environmental Monitoring. 2006;**8**:456–464. DOI: 10.1039/b517512g

[26] Magnusson M, Heimann K, Negri AP. Comparative effects of herbicides on photosynthesis and growth of tropical estuarine nicroalgae. Marine Pollution Bulletin. 2008;**56**:1545–1552. DOI: 10.1016/j.marpolbul.2008.05.023

[27] Bengtson Nash SM, McMahon K, G. Eaglesham G, Müller JF. Application of a novel phytotoxicity assay for the detection of herbicides in Hervey Bay and the Great Sandy Straits. Marine Pollution Bulletin. 2005;**51**:351–360. DOI:10.1016/j.marpolbul.2004.10.017

[28] Sargian P, Pelletier É, Mostajir B, Ferreyra GA, Demers S. TBT toxicity on a natural planktonic assemblage exposed to enhanced ultraviolet-B radiation. Aquatic Toxicology. 2005;**73(3)**:299–314. DOI: 10.1016/j.aquatox.2005.03.019

[29] Mallick N, Mohn FH. Use of chlorophyll fluorescence in metal stress research; a case study with the green microalga *Scenedesmus*. Ecotoxicology and Environmental Safety. 2003;**55**:64–69. DOI: 10.1016/S0147-6513(02)00122-7

[30] Carrera-Martinez D, Mateos-Sanz A, Lopez-Rodas V, Costas E. Adaptation of microalgae to a gradient of continuous petroleum contamination. Aquatic Toxicology. 2011;**101**:342–350. DOI: 10.1016/j.aquatox.2010.11.009

[31] Walters CC. The origin of petroleum. In: Practical Advances in Petroleum Processing, Vol. 1. New York: Springer Science and Business Media Inc.; 2007. pp. 79–101. DOI: 10.1007/978-0-387-25789-1_2

[32] Lewis M, Pryor R. Toxicities of oils, dispersants and dispersed oils to algae and aquatic plants: Review and database value to resource sustainability. Environmental Pollution. 2013;**180**:345–367. DOI: 10.1016/j.envpol.2013.05.001

[33] Saadoun IMK. Impact of oil spills on marine life. In: Larramendy M, Soloneski S, editors. Emerging Pollutants in the Environment – Current and Further Implications. InTech; 2015: Chapter 4. ISBN 978-953-51-2160-2. DOI: 10.5772/60455.

[34] Tjeerdema R, Bejarano AC, Edge S. Biological effects of dispersants and dispersed oil on surface and deep ocean species. From the Oil Spill Dispersant-related Research Workshop, Hosted by the Center for Spills in the Environment, March 12–13, 2013, Baton Rouge, LA, United States of America. Available from: https://crrc.unh.edu/sites/crrc.unh.edu/files/tjeerdemabejeranoedge.pdf [Assessed: 2017-01-06]

[35] González J, Figueiras FG, Aranguren-Gassis M, Crespo BG, Fernandez E, Moran XAG, Nieto-Cid M. Effect of a simulated oil spill on natural assemblages of marine phytoplankton enclosed in microcosms. Estuarine, Coastal and Shelf Science. 2009;**83(3)**:265–276. DOI: 10.1016/j.ecss.2009.04.001

[36] Batterton JC, Winters K, Van Baalen C. Sensitivity of three microalgae to crude oils and fuel oils. Marine Environmental Research. 1978;**1(1)**:31–41. DOI: 10.1016/0141-1136(78)90011-9

[37] Hsiao SIC, Kittle DW, Foy MG. Effects of crude oils and the oil dispersant Corexit on primary production of Arctic marine phytoplankton and seaweed. Environmental Pollution. 1978;**15**:209–221. DOI: 10.1016/0013-9327(78)90066-6

[38] Giddings JM, Washington JN. Coal-liquefaction products, shale oil and petroleum, acute toxicity to freshwater algae. Environmental Science and Technology. 1981;**15(1)**:106–108. DOI: 10.1021/es00083a013

[39] Karydis M. The toxicity of crude oil for the marine alga *Skeletonema costatum* (Greville) Cleve in relation to nutrient limitation. Hydrobiologia. 1981;**85**:137–143. DOI: 10.1007/BF00006623

[40] Ostgaard K, Eide I, Jensen A. Exposure of phytoplankton to Ecofisk oil. Marine Environmental Research. 1984;**11(3)**:183–200. DOI: 10.1016/0141-1136(84)90045-X

[41] Chan K, Chiu SY. The effects of diesel oil and oil dispersants on growth, photosynthesis and respiration of *Chlorella salina*. Archives Environmental Contamination and Toxicology. 1985;**14**:325–331. DOI: 10.1007/BF01055410

[42] Hegseth EN, Ostgaard K. Application of in-situ dialysis cultures in studies of phytotoxicity of North Sea crude oils. Water Research, 1985;**19(3)**:383–391. DOI: 10.1016/0043-1354(85)90099-5

[43] Siron R, Pelletier E, Delille D, Roy S. Fate and effects of dispersed crude oil under icy conditions simulated in mesocosms. Marine Environmental Research. 1993;**35**:273–302. DOI: 10.1016/0141-1136(93)90098-K

[44] El-Dib MA, Abou-Waly, HF. Impact of fuel oil on the freshwater alga *Selenastrum capricornutum*. Bulletin Environmental Contamination Toxicology. 1997;**59**:438–444. DOI: 10.1007/s001289900497

[45] Huang YJ, Jiang BJ, Zeng JN, Chen QZ, Zhao YQ, Liao YB, Shou L, Xu XQ. 2010. The chronic effects of oil pollution on marine phytoplankton in a subtropical bay, China. Environmental Environmental Monitoring and Assessment. 2010;**176**:517–530. DOI: 10.1007/s10661-010-1601-6

[46] Jaiswar JRM, Ram A, Rokade MA, Karangutkar S, Yengal B, Dalvi S, Acharya D, Sharma S, Gajbhiye SN. Phytoplankton dynamic responses to oil spill in Mumbai Harbour. International Journal of Innovative Biological Research. 2013;**2(1)**:30–50. DOI: 10.2169/ijibr.2013.2.1.4

[47] Tewari A, Joshi HV, Trivedi RH, Sravankumar VG, Raghunathan C, Khambhaty Y, Kotiwar OS, Mandal SK. 2001. The effect of ship scrapping industry and its associated wastes on the biomass production and biodiversity of biota in in situ condition at Alang. Marine Pollution Bulletin. 2001;**42**:462–469. Available from: https://www.ncbi.nlm.nih.gov/pubmed/11468924 [Accessed: 2017-01-10]

[48] Salas N, Ortiz L, Gilcoto M, Varela M, Bayona JM, Groom S, Alvarez-Salgado XA, Albaiges J. Fingerprinting petroleum hydrocarbons in plankton and surface sediments during the spring and early summer blooms in the Galician coast (NW Spain) after the Prestige oil spill. Marine Environmental Research. 2006;**62**:388–413. DOI: 10.1016/j.marenvres.2006.06.004

[49] Al Obaidy AHMJ, Lami MHM. The toxic effects of crude oil in some freshwater Cyanobacteria. Journal of Environmental Protection. 2014;**5**:359–367.Available from: http://dx.doi.org/10.4236/jep.2014.55039 [Assessed: 2017-01-06]

[50] Gaur JP, Kumar HD. Growth response of four micro-algae to three crude oils and a furnace oil. Environmental Pollution A. 1981;**25**:77–85. DOI: 10.1016/0143-1471(81)90116-1

[51] Ostgaard K, Aaberg A, Klungsoyr J, Jensen A. Comparative studies of phytotoxicity and chemical composition of aqueous oil solutions affected by evaporation, illumination and extraction. Water Research. 1987;**21(2)**:155–164. DOI: 10.1016/0043-1354(87)90044-3

[52] Shailaja MS. The influence of dissolved petroleum hydrocarbon residues on natural phytoplankton biomass. Marine Environmental Research. 1988;**25**:315–324. DOI: 10.1016/0141-1136(88)90018-9

[53] Gaur JP, Singh K. Comparative studies on the toxicity of petroleum oils and their aqueous extracts towards *Anabaena doliolum*. Proceedings of the Indian Academy of Science. 1989;**99**:459–466. DOI: 10.1007/BF03053615

[54] Sargian P, Mostajir B, Chatila K, Ferreyra GA, Pelletier E, Demers S. Non-synergistic effects of water-soluble crude oil and enhanced ultraviolet-B radiation on a natural plankton assemblage. Marine Ecology Progress Series. 2005;**294**:63–77. Available from: http://www.int-res.com/articles/meps2005/294/m294p063.pdf [Accessed: 2017-01-19]

[55] Lee CI. Temporal variation of chlorophyll a concentration in the coastal waters affected by the Hebei Spirit oil spill in the West Sea of Korea. Marine Pollution Bulletin. 2009;**58**:496–502. DOI: 10.1016/j.marpolbul.2008.12.007

[56] Scott BF, Glooschenko V. Impact of oil and oil-dispersant mixtures on flora and water chemistry parameters in freshwater ponds. Science of the Total Environment. 1984;**35**:169–190. DOI: 10.1016/0048-9697(84)90061-5

[57] Brown HM, Goudey JS, Foght JM, Cheng SK, Dale M, Hoddinolt J, Quaife LR, Westlake DWS. Dispersion of spilled oil in freshwater systems: field trial of a chemical dispersant. Oil and Chemical Pollution. 1990;**6(1)**:37–54. DOI: 10.1016/S0269-8579(05)80038-3

[58] Marshall PA, Edgar, GJ. The effect of the Jessica grounding on subtidal invertebrate and plant communities at the Galapagos wreck site. Marine Pollution Bulletin. 2003;**47**:284–295. DOI: 10.1016/S0025-326X(03)00157-7

[59] Peterson CH, Rice SD, Short JW, Esler D, Bodkin JL, Ballachey BE, Irons DB. Long-term ecosystem response to the Exxon Valdez oil spill. Science. 2003;**302**:2082–2086. DOI: 10.1126/science.1084282

[60] Carrera-Martinez D, Mateos-Sanz A, Lopez-Rodas V, Costas E. Microalgae response to petroleum spill: an experimental model analyzing physiological and genetic response of *Dunaliella tertiolecta* (Chlorophyceae) to oil samples from the tanker *Prestige*. Aquatic Toxicology. 2010;**97**:151–157. DOI: 10.1016/j.aquatox.2009.12.016

[61] Varela M, Bode A, Lorenzo J, Alvarez-Ossorio MT, Miranda A, Patrocinio T, Anadon R, Viesca L, Rodriguez N, Valdes L, Cabal J, Urrutia A, Garcia-Soto C, Rodriguez M, Alvarez-Salgado XA, Groom S. The effect of the "Prestige" oil spill on the plankton of the N-NW Spanish coast. Marine Pollution Bulletin. 2006;**53**:272–286. DOI: 10.1016/j.marpolbul.2005.10.005

[62] Gin KYH, Huda MDK, Lim WK, Tkalich P. An oil spill–food interaction model for coastal waters. Marine Pollution Bulletin. 2001;**42**:590–597. DOI:10.1016/S0025-326X(00)00205-8

[63] Singh AK, Gaur JP. Effects of petroleum oils and their paraffinic, asphaltic, and aromatic fractions on photosynthesis and respiration of microalgae. Ecotoxicology and Environmental Safety. 1990;**19(1)**:8–16. DOI: 10.1016/0147-6513(90)90073-E

[64] El-Sheekh MM, El-Naggar AH, Osman MEH, HaiederA. 2000. Comparative studies on the green algae *Chlorella homosphoera* and *Chlorella vulgaris* with respect to oil

pollution in the River Nile. Water Air and Soil Pollution. 2000;**124**:187–204. DOI: 10.1023/A:1005268615405

[65] Carter AD. Herbicide movement in soils: principles, pathways and processes. Weed Research. 2000;**40 (1)**:113–122.DOI: 10.1046/j.1365-3180.2000.00157.x

[66] Pehkonen SO, Zhang Q. The degradation of organophosphorus pesticides in natural waters: a critical review. Critical Reviews in Environmental Science and Technology. 2002;**32(1)**:17–72. DOI: 10.1080/10643380290813444

[67] Sherman TD, Vaughn KC, Duke SO. Mechanisms of action and resistance to herbicides. In: Duke SO, editor. Herbicides-resistant Crops: Agricultural, Environmental, Economic, Regulatory, and Technical Aspects. 1st ed. CRC Press, Boca Raton, Florida. 1996. pp. 13–35.

[68] Rutherford AW, Krieger-Liszkay A. Herbicide-induced oxidative stress in photosystem II. Trends in Biochemical Sciences. 2001;**26(11)**:648–653. DOI: 10.1016/S0968-0004 (01)01953-3

[69] Couderchet M, Vernet, G. Pigments as biomarkers of exposure to the vineyard herbicide flazasulfuron in freshwater algae. Ecotoxicology and Environmental Safety. 2003;**55(3)**:271–277. DOI: 10.1016/S0147-6513(02)00064-7

[70] Sikka HC, Pramer D. (1968). Physiological effects of fluometuron on some unicellular algae. Weed Science. 1968;**16(3)**:296–299. Available from http://www.jstor.org/stable/404132 [Accessed: 2017-01-10]

[71] DeLorenzo ME, Leatherbury M, Weiner JA, Lewitus AJ, Fulton MH. Physiological factors contributing to the species-specific sensitivity of four estuarine microalgal species exposed to the herbicide atrazine. Aquatic Ecosystem Health and Management. 2004;**7**:137–146. DOI: 10.1080/14634980490281551

[72] Tang J, Hoagland KD, Siegfried BD. Uptake and bioconcentration of atrazine by selected freshwater algae. Environmental Toxicology and Chemistry. 1998;**17(6)**:1085–1090. DOI: 10.1002/etc.5620170614

[73] Bérard A, Dorigo U, Mercier I, Beker van Slooten K, Grandjean D, Leboulanger C. Comparison of the ecotoxicological impact of the triazines Irgarol 1051 and atrazine on microalgal cultures and natural microalgal communities in Lake Geneva. Chemosphere. 2003;**53**:935–944. DOI: 10.1016/S0045-6535(03)00674-X

[74] Guasch H, Sabater S. Light history influences the sensitivity to atrazine in periphytic algae. Journal of Phycology. 1998;**34**:233–241. DOI: 10.1046/j.1529-8817.1998.340233.x

[75] Navarro E, Guasch H, Sabater S. Use of microbenthic algal communities in ecotoxicological tests for the assessment of water quality: The Ter river case study. Journal of Applied Phycology. 2002;**14(1)**:41–48. DOI: 10.1023/A:1015242301451

[76] Debenest T, Silvestre J, Coste M, Pinelli E. Effects of pesticides on freshwater diatoms. Vol. 203 of the series. Reviews of Environmental Contamination and Toxicology. 2009;203:87–103. Available from: http://link.springer.com/chapter/10.1007%2F978-1-4419-1352-4_2 [Accessed: 2017-01-20]

[77] Goldsborough LG, Brown DJ. Effect of glyphosate (RoundupR formulation) on periphytic algal photosynthesis. Bulletin of Environmental Contamination and Toxicology. 1998;**41(2)**:253–260. DOI: 10.1007/BF01705439

[78] Weiner JA, DeLorenzo ME, Fulton MH. Atrazine induced species-specific alterations in the subcellular content of microalgal cells. Pesticide Biochemistry and Physiology. 2007;**87(1)**:47–53. DOI: 10.1016/j.pestbp.2006.06.003

[79] Li X, Ping X, Xiumei S, Zhenbin W, Liqiang X. Toxicity of cypermethrin on growth, pigments and superoxide dismutase of *Scenedesmus obliquus*. Ecotoxicology and Environmental Safety. 2005;**60(2)**:188–192. DOI: 10.1016/j.ecoenv.2004.01.012

[80] DeLorenzo ME, Scott GI, Ross PE. Toxicity of pesticides to aquatic microorganisms: a review. Environmental Toxicology and Chemistry. 2001;**20(1)**:84–98. DOI: 10.1002/etc.5620200108

[81] Snel JFH, Vos JH, Gylstra R, Brock TCM. Inhibition of photosystem II (PSII) electron transport as a convenient endpoint to assess stress of the herbicide linuron on freshwater plants. Aquatic Ecology. 1998;**32(2)**:113–123. DOI: 10.1023/A:1009971930626

[82] Arrhenius A, Gronvall F, Scholze M, Backhaus T, Blanck H. Predictability of the mixture toxicity of 12 similarly acting congeneric inhibitors of photosystem II in marine periphyton and epipsammon communities. Aquatic Toxicology. 2004;**68**:351–367. DOI: 10.1016/j.aquatox.2004.04.002

[83] Knauert S, Escher B, Singer H, Hollender J, Knauer K. Mixture toxicity of three photosystem II inhibitors (atrazine, isoproturon, and diuron) toward photosynthesis of freshwater phytoplankton studied in outdoor mesocosms. Environmental Science Technology. 2008;**42**:6424–6430. DOI: 10.1021/es072037q

[84] Zhang YH, Liu SS, Song XQ, Ge HL. Prediction for the mixture toxicity of six organophosphorus pesticides to the luminescent bacterium *Q67*. Ecotoxicology and Environmental Safety. 2008;**71(3)**:880–888. DOI: 10.1016/j.ecoenv.2008.01.014

[85] Altenburger R, Backhaus T, Boedeker W, Faust M, Scholze M, Grimme LH. Predictability of the toxicity of multiple chemical mixtures to *Vibrio fischeri*: mixtures composed of similarly acting chemicals. Environmental Toxicology and Chemistry. 2000;**19(9)**:2341–2347. DOI: 10.1002/etc.5620190926

[86] Faust M, Altenburger R, Backhaus T, Blanck H, Boedeker W, Gramatica P, Hamer V, Scholze M, Vighi M, Grimme L.H. Predicting the joint algal toxicity of multi-component s-triazine mixtures at low-effect concentrations of individual toxicants. Aquatic Toxicology. 2001;**56**:13–32. DOI: 10.1016/S0166-445X(01)00187-4

[87] Kim Tiam S, Fauvelle V, Morin S, Mazzella N. Improving toxicity assessment of pesticide mixtures: the use of polar passive sampling devices extracts in microalgae toxicity tests. Frontiers in Microbiology. 2016;**7**:1388. DOI:10.3389/fmicb.2016.01388

[88] Ma J, Wang S, Wang P, Ma l, Chen X, Xu R. Toxicity assessment of 40 herbicides to the green alga *Raphidocelis subcapitata*. Ecotoxicology and Environmental Safety. 2006;**63**:456–462. DOI: 10.1016/j.ecoenv.2004.12.001

[89] Tlili A, Montuelle B, Bérard A, Bouchez A. Impact of chronic and acute pesticide exposures on periphyton communities. Science of the Total Environment. 2011;**409**:2102–2113. DOI:10.1016/j.scitotenv.2011.01.056

[90] Backhaus T, Altenburger R, Boedeker W, Faust M, Scholze M, Grimme LH. Predictability of the toxicity of multiple mixture of dissimilarly acting chemicals to *Vibrio Fischeri*. Environmental Toxicology and Chemistry. 2000;**19(9)**:2348–2356. DOI: 10.1002/etc.5620190927

[91] Backhaus T, Arrhenius A, Blanck H. Toxicity of a mixture of dissimilarly acting substances to natural algal communities: predictive power and limitations of independent action and concentration addition. Environmental Science Technology. 2004;**38**:6363–6370. DOI: 10.1021/es0497678

[92] Fairchild JF, Lapoint TW, Schwartz TR. Effects of an herbicide and insecticide mixture in aquatic mesocosms. Achieves in Environmental Toxicology and Chemistry. 1994;**27**:527–533. DOI: 10.1007/BF00214845

[93] Hartgers EM, Aalderink GHR, van den Brink PJ, Gylstra R, Wiegman JWF, Brock TCM. Ecotoxicological threshold levels of a mixture of herbicides (atrazine, diuron and metolachlor) in freshwater microcosms. Aquatic Ecology. 1998;**32**:135–152. DOI: 10.1023/A:1009968112009

[94] van Wijngaarden RPA, Cuppen JGM, Arts GHP, Crum SJH, van den Hoorn MW, van den Brink PJ, Brock TCM. Aquatic risk assessment of a realistic exposure to pesticides used in bulb crops: a microcosm study. Environmental Toxicology and Chemistry. 2004;**23**:1479–1498. DOI: 10.1897/03-80

[95] Hoagland KD, Drenner RW, Smith JD, Cross DR. Fresh-water community responses to mixtures of agricultural pesticides - Effects of atrazine and bifenthrin. Environmental Toxicology and Chemistry.1993;**12**:627–637. DOI: 10.1002/etc.5620120404

[96] Deneer JW. Toxicity of mixtures of pesticides in aquatic systems. Pest Management Science. 2000;**56**:516–520. DOI: 10.1002/(SICI)1526-4998(200006)56:6<516::AID-PS163>3.0.CO;2–0

[97] Lydy M, Belden J, Wheelock C, Hammock B, Denton D. Challenges in regulating pesticide mixtures. Ecology and Society. 2004;**9(6)**:1. Available from: http://www.ecologyandsociety.org/vol9/iss6/art1. [Accessed: 2016-12-28]

[98] Belden JB, Gilliom RJ, Lydy MJ. How well can we predict the toxicity of pesticide mixtures to aquatic life? Integrated Environmental Assessment and Management. 2007;**3(3)**:364–372. DOI: 10.1002/ieam.5630030307

[99] Cedergreen N, Christensen AM, Kamper A, Kudsk P, Mathiassen SK, Streibig JC, Sørensen H. A review of independent action as a reference model for binary mixtures of compounds with different molecular target sites. Environmental Toxicology and Chemistry. 2008;**27(7)**:1621–1632. DOI: 10.1897/07-474.1

[100] Boobis A, Budinsky R, Collie S, Crofton K, Embry M, Felter S, Hertzberg R, Kopp D, Mihlan G, Mumzat M, Price P, Solomon K, Teuschler L, Yang R, Zaleski R. Critical analysis of literature on low-dose synergy for use in screening chemical mixtures for risk assessment. Critical Reviews in Toxicology. 2011;**41(5)**:369–383. DOI: 10.3109/10408444.2010.543655

[101] Cedergreen, N. Quantifying synergy: a systematic review of mixture toxicity studies within environmental toxicology. PLoS ONE. 2014;**9(5)**:e96580. DOI: 10.1371/journal.pone.0096580

[102] Konstantinou IK, Albanis TA. Worldwide occurrence and effects of antifouling paint booster biocides in the aquatic environment: a review. Environmental International. 2004;**30(2)**:235–248. DOI: 10.1016/S0160-4120(03)00176-4

[103] Evans SM, Leksono T, McKinnell PD. Tributyltin pollution: a diminishing problem following legislation limiting the use of TBT-based anti-fouling paints. Marine Pollution Bulletin. 1995;**30(1)**:14–21. DOI: 10.1016/0025-326X(94)00181-8

[104] Fent K. Ecotoxicology of organotin compounds. Critical Reviews in Toxicology. 1996;**26(1)**:3–117. DOI: 10.3109/10408449609089891

[105] Yoo YH, Sidharthan M, Shin HW. 2007. Effects of tributyl-tin on a marine microalga, *Tetraselmis suecica*. Journal of Environmental Biology. 2007;**28 (3)**:571–575. Available from: http://www.jeb.co.in/journal_issues/200707_jul07/paper_08.pdf [Accessed: 2016-12-29]

[106] Rumampuk NDC, Rumengan IFM, Ohji M, Arai T, Nobuyuki M. Effects of tributyltin on the chlorophyll contents of marine microalga *Tetraselmis tetrathele, Nannochloropsis oculata* and *Dunaliella* sp. Coastal Marine Science. 2004;**29(1)**: 40–44. Available from: http://ci.nii.ac.jp/naid/110004302117/ [Accessed: 2016-12-29]

[107] Sidharthan M, Young KS, Woul LH, Soon PK, Shin HW. 2002. TBT toxicity on the marine microalga *Nannochloropsis oculata*. Marine Pollution Bulletin. 2002;**45(1–12)**:177–180. DOI: 10.1016/S0025-326X(01)00283-1

[108] Jensen HF, Holmer M, Dahllöf I. Effects of tributyltin (TBT) on the seagrass Ruppia maritima. Marine Pollution Bulletin. 2004;**49(7–8)**:564–573. DOI: 10.1016/j.marpolbul.2004.03.010

[109] Hall LW, Gardinali P. Ecological risk assessment for Irgarol 1051 and its major metabolite in United States surface waters. Human and Ecological Risk Assessment. 2004;**10(3)**:525–542. DOI: 10.1080/10807030490452188

[110] Hall LW, Giddings JM, Solomon KR, Balcomb R. An ecological risk assessment for the use of Irgarol 1051 as an algaecide for antifoulant paints. Critical Reviews in Toxicology.

1999;**29(4)**:367–437. Available from: https://www.ncbi.nlm.nih.gov/pubmed/10451264 [Accessed: 2016-12-30]

[111] Hall LW, Killen WD, Anderson RD, Gardinali PR, Balcomb R. Monitoring of Irgarol 1051 concentrations with concurrent phytoplankton evaluations in East Coast areas of the United States. Marine Pollution Bulletin. 2005;**50(6)**:668–681. DOI: 10.1016/j.marpolbul.2005.02.025

[112] Owen R, Knap A, Toaspern M, Carbery K. Inhibition of coral photosynthesis by the antifouling herbicide Irgarol 1051. Marine Pollution Bulletin. 2002;**44(7)**:623–632. DOI: 10.1016/S0025-326X(01)00303-4

[113] Guardiola FA, Cuesta A, Meseguer J, Esteban MA. Risks of using antifouling biocides in aquaculture. International Journal of Molecular Sciences. 2012;**13(2)**:1541–1560. DOI: 10.3390/ijms13021541

[114] Dahl B, Blanck H. Toxic effects of the antifouling agent Irgarol 1051 on periphyton communities in coastal water microcosms. Marine Pollution Bulletin. 1996;**32(4)**:342–350. DOI: 10.1016/0025-326X(96)84828-4

[115] Devilla RA, Brown MT, Donkin M, Tarran GA, Aiken J, Readman JW. Impact of antifouling booster biocides on single microalgal species and on a natural marine phytoplankton community. Inter-Research Marine Ecology Progress Series. 2005;**286**:1–12. DOI: 10.3354/meps286001

[116] Tolhurst LE, Barry J, Dyer RA, Thomas KV. The effect of resuspending sediment contaminated with antifouling paint particles containing Irgarol 1051 on the marine macrophyte *Ulva intestinalis*. Chemosphere. 2007;**68(8)**:1519–1524. DOI: 10.1016/j.chemosphere.2007.03.005

[117] Kottuparambil S, Lee S, Han T. Single and interactive effects of the antifouling booster herbicides diuron and irgarol 1051 on photosynthesis in the marine cyanobacterium, *Arthrospira maxima*. Toxicology and Environmental Health Sciences. 2013;**5(2)**:71–81. DOI: 10.1007/s13530-013-0157-6

[118] Nyström B, Slooten BCK, Berard A, Grandjean D, Druart JC, Leboulanger C. Toxic effects of Irgarol 1051 on phytoplankton and macrophytes in Lake Geneva. Water Research. 2002;**36(8)**:2020–2028. DOI: 10.1016/S0043-1354(01)00404-3

[119] Buma AGJ, Sjollema SB, van de Poll WH, Klamer HJC, Bakker JF. Impact of the antifouling agent Irgarol 1051 on marine phytoplankton species. Journal of Sea Research. 2009;**61**:133–139. DOI: 10.1016/j.seares.2008.11.007

[120] Seery CR, Gunthorpe L, Ralph, PJ. Herbicide impact on *Hormosira banksia* gametes measured by fluorescence and germination bioassays. Environmental Pollution. 2006;**140(1)**:43–51. DOI: 10.1016/j.envpol.2005.07.001

[121] Scarlett A, Donkin ME, Fileman TW, Donkin P. Occurrence of the marine antifouling agent Irgarol 1051 within the Plymouth sound locality: implications for the green

macroalga *Enteromorpha intestinalis*. Marine Pollution Bulletin. 1997;**34(8)**:645–651. DOI: 10.1016/S0025-326X(96)00187-7

[122] Ricart M, Barcelo D, Geiszinger A, Guasch H, Lopez De Alda M, Romani A, Romaní AM, Vidal G,Villagrasa M, Sergi Sabater S. Effects of low concentrations of the phenylurea herbicide diuron on biofilm algae and bacteria. Chemosphere. 2009;**76(10)**:1392–1401. DOI: 10.1016/j.chemosphere.2009.06.017

[123] Tlili A, Dorigo U, Montuelle B, Margoum C, Carluer N, Gouy V, Bouchez A, Bérard A. Responses of chronically contaminated biofilms to short pulses of diuron—an experimental study simulating flooding events in a small river. Aquatic Toxicology. 2008;**87(4)**:252–263. DOI: 10.1016/j.aquatox.2008.02.004

[124] Collier PJ, Austin P, Gilbert P. Isothiazolone biocides: enzyme inhibition pro-drugs. International Journal of Pharmaceuticals. 1991;**74**:195–201. DOI: 10.1016/0378-5173(91)90237-I

[125] Duffus JH. Heavy metal – a meaningless term? (IUPAC Technical Report). Pure and Applied Chemistry. 2002;**74(5)**:793–807. DOI: 10.1351/pac200274050793

[126] Arunakumara KKIU, Zhang X. Heavy metal bioaccumulation and toxicity with special reference to microalgae. Journal of Ocean University China. 2008;**7(1)**:60–64. DOI: 10.1007/s11802-008-0060-y

[127] Nieboer E, Richardson DHS. The replacement of the nondescript term "heavy metals" by the biologically and chemically significant classification of metal ions. Environmental Pollution, Series B. 1980;**B1**:3–26. Available from: http://citeseerx.ist.psu.edu/viewdoc/download?doi=10.1.1.472.7308&rep=rep1&type=pdf [Accessed: 2017-01-07]

[128] Valavanidis A, Vlachogianni T. Metal pollution in ecosystems. Ecotoxicology studies and risk assessment in the marine environment. Science Advances on Environment, Toxicology and Ecotoxicology Issues [Internet]. 2010. Available from: http://chem-tox-ecotox.org/wp/?p=394 [Accessed: 2016-12-21]

[129] Zhou Q, Zhang J, Fu J, Shi J, Jiang G. Biomonitoring: an appealing tool for assessment of metal pollution in the aquatic ecosystem. Analytica Chimica Acta. 2008;**606**:135–150. DOI: 10.1016/j.aca.2007.11.018

[130] Rai PK. Heavy-metal pollution in aquatic ecosystems and its phytoremediation using wetland plants: an eco-sustainable approach. International Journal of Phytoremediation. 2008;**10(2)**:133–160. DOI: 10.1080/15226510801913918

[131] Rai PK. Heavy metal pollution in lentic ecosystem of sub-tropical industrial region and its phytoremediation. International Journal of Phytoremediation. 2010;**12(3)**:226–242. DOI: 10.1080/15226510903563843

[132] Appenroth K. Definition of "heavy metals" and their role in biological systems. In: Sherameti I, Varma A. editors. Soil Heavy Metals, Soil Biology. Berlin, Heidelberg: Springer-Verlag; 2010. pp. 19–29. ISBN: 978-3-642-02435-1

[133] Bakar LA, Amina NM, Zakeria HM. Impact of mercury (II) nitrate on physiological and biochemical characteristics of selected marine algae of different classes. Procedia Environmental Sciences. 2015;**30**:222–227. DOI: 10.1016/j.proenv.2015.10.040

[134] Rodríguez MC, Barsanti L, Passarelli V, Evangelista V, Conforti V, Gualtieri P. Effects of chromium on the photosynthetic and photoreceptive apparatus of the alga *Chlamydomonas reinhardtii*. Environmental Research. 2007;**105(2)**:234–239. DOI: 10.1016/j.envres.2007.01.011

[135] Singh S, Parihar P, Singh R, Singh VP, Prasad SM. Heavy metal tolerance in plants: role of transcriptomics, proteomics, metabolomics, and ionomics. Frontiers in Plant Science. 2015;**6**:1143. DOI: 10.3389/fpls.2015.01143

[136] Mysliwa-Kurdziel- B, Prasad MNV, Stralka K. Photosynthesis in heavy metal stress plants. In: Prasad MNV, editor. Heavy Metal Stress in Plants. 3rd ed. Berlin: Springer; 2004. pp. 146–181. ISBN: 978-3-662-07743-6

[137] Barcelo J, Poschnrieder C. Structural and ultrastructural changes in heavy metal exposed plants. In: Prasad MNV, editor. Heavy Metal Stress in Plants. 3rd ed. Berlin: Springer; 2004. pp. 223–248. ISBN: 978-3-662-07743-6

[138] De Filippis LF, Hampp R, Ziegler H. The effects of sublethal concentrations of zinc, cadmium and mercury on *Euglena*. II. Respiration, photosynthesis and photochemical activities. Archives of Microbiology. 1981;**128(4)**:407–411. DOI: 10.1007/BF00405922

[139] Demidchik V, Sokolik A, Yurin V. Characteristics of non-specific permeability and H^+-ATPase inhibition induced in the plasma membrane of *Nitella flexilis* by excessive Cu^{2+}. Planta. 2001;**212(4)**:583–590. DOI: 10.1007/s004250000422

[140] Clijsters H, Van Assche F. Inhibition of photosynthesis by heavy metals. Photosynthesis Research. 1985;**7(1)**:31–40. DOI: 10.1007/BF00032920

[141] Kupper H, Setlik I, Spiller M, Kupper FC, Prasil O. Heavy metal induced inhibition of photosynthesis: targets of in vivo heavy metal chlorophyll formation. Journal of Phycology. 2002;**38**:429–441. DOI:10.1046/j.1529-8817.2002.t01-1-01148.x

[142] Shi X, Dalal NS. On the hydroxyl radical formation in the reaction between hydrogen peroxide and biologically generated chromium (V) species. Archives of Biochemistry and Biophysics. 1990;**277(2)**:342–350. DOI: 10.1016/0003-9861(90)90589-Q

[143] Kukarskikh GL, Graevskaia EE, Krendeleva TE, Timofeedv KN, Rubin AB. Effect of methylmercury on primary photosynthesis processes in green microalgae *Chlamydomonas reinhardtii*. Biofizika. 2003;**48(5)**:853–859. Available from: http://europepmc.org/abstract/MED/14582410 [Accessed: 2017-01-01]

[144] Nicolardi V, Cai G, Parrotta L, Puglia M, Bianchi L, Bini L, Gaggi C. The adaptive response of lichens to mercury exposure involves changes in the photosynthetic machinery. Environmental Pollution. 2012;**160**:1–10. DOI: 10.1016/j.envpol.2011.09.015

[145] Miao AJ, Wang WX, Juneau P. Comparison of Cd, Cu, and Zn toxic effects on four marine phytoplankton by pulse-amplitude-modulated fluometry. Environmental Toxicology and Chemistry. 2005;**24(10)**:2603–2011. DOI: 10.1897/05-009R.1

[146] Bohner H, Bohme H, Boger P. Reciprocal formation of plastocyanin and cytochrome c-553 and the influence of cupric ions on photosynthetic electron transport. Biochimica et Biophysica Acta. 1980;**592**:103–112. Available from: https://www.ncbi.nlm.nih.gov/pubmed/6249351 [Accessed: 2017-01-01]

[147] Cid A, Herrero C, Torres E, Abalde J. Copper toxicity on the marine microalga Phae*odactylum tricornutum*: effects on photosynthesis and related parameters. Aquatic Toxicology. 1995;**31**:165– 174. DOI: 10.1016/0166-445X(94)00071-W

[148] Jegerschold C, Arellano JB, Schroder WP, Van Kan PJM, Baron M, Styring S. Copper (II) inhibition of electron transfer through photosystem II studied by EPR spectroscopy. Biochemistry. 1995;**34**:12747–12754. DOI: 10.1021/bi00039a034

[149] Einicker-Lamas M, Antunes Mezian G, Benavides Fernandes T, Silva FLS, Guerra F, Miranda K, Attia, M, Oliveira MM, 2002. *Euglena gracilis* as a model for study of Cu^{2+} and Zn^{2+} toxicity and accumulation in eukaryotic cells. Environmental Pollution. 2002;**120**:779–786. DOI: 10.1016/S0269-7491(02)00170-7

[150] Rocchetta I, Kupper H. Chromium- and copper-induced inhibition of photosynthesis in Euglena gracilis analysed on the single-cell level by fluorescence kinetic microscopy. New Phytologist. 2009;**182(2)**:405–420. DOI: 10.1111/j.1469-8137.2009.02768.x

[151] Dewez D, Geoffroy L, Vernet G, Popovic R. 2005. Determination of photosynthetic and enzymatic biomarkers sensitivity used to evaluate toxic effects of copper and fludioxonil in alga *Scenedesmus obliquus*. Aquatic Toxicology. 2005;**74(2)**:150–159. DOI: 10.1016/j.aquatox.2005.05.007

[152] Perales-Vela HV, Gonzalez-Moreno S, Montes-Horcasitas C, CanizaresVillanueva RO. Growth, photosynthetic and respiratory responses to sub-lethal copper concentrations in *Scenedesmus incrassatulus* (Chlorophyceae). Chemosphere. 2007;**67**:2274–2281. DOI: 10.1016/j.chemosphere.2006.11.036

[153] Xia J, Tian Q. Early stage toxicity of excess copper to photosystem II of *Chlorella pyrenoidosa*–OJIP chlorophyll a fluorescence analysis. Journal of Environmental Sciences. 2009;**21**:1569–1574. DOI: 10.1016/S1001-0742(08)62457-2

[154] Ou-Yang HL, Kong XZ, He W, Qin N, He QS, Yan W, Rong W, Liu XF. Effects of five heavy metals at sub-lethal concentrations on the growth and photosynthesis of *Chlorella vulgaris*. Chinese Science Bulletin. 2012;**57**:3363–3370. DOI: 10.1007/s11434-012-5366-x

[155] Sbihi K, Cherifi O, El gharmali A, Oudra B, Aziz F. Accumulation and toxicological effects of cadmium, copper and zinc on the growth and photosynthesis of the freshwater diatom *Planothidium lanceolatum* (Brébisson) Lange-Bertalot: a laboratory study. Journal of Materials and Environmental Science. 2012;**3(3):**497–506. Available from:

http://www.jmaterenvironsci.com/Document/vol3/vol3_N3/49-JMES-178-2012-Sbihi.pdf [Accessed: 2017-01-21]

[156] El-Sheekh MM, El-Naggaar AH, Osman MEH, El-Mazaly E. Effect of cobalt on growth, pigments and the photosynthetic electron transport in *Monoraphidium minutum* and *Nitzchia perminuta*. Brazilian Journal of Plant Physiology. 2003;**15(3)**:159–166. DOI: 10.1590/S1677-04202003000300005

[157] Hiriat-Baer VP, Fortin C, Lee DY, Campbell PG. Toxicity of silver to two freshwater algae, *Chlamydomonas reinhardtii* and *Pseudokirchneriella sub-capitata*, grown under continuous culture conditions: influence of thiosulphate. Aquatic Toxicology. 2006;**78(2)**:136–148. DOI: 10.1016/j.aquatox.2006.02.027

[158] Juarez AB, Barsanti L, Passarelli V, Evangelista V, Vesentini N, Conforti V, Gualtieri P. *In vivo* microspectroscopy monitoring of chromium effects on the photosynthetic and photoreceptive apparatus of *Eudorina unicocca* and *Chlorella kessleri*. Journal of Environmental Monitoring. 2008;**10**:1313–1318. DOI: 10.1039/B809566C

[159] Navarro E, Piccapietra F, Wagner B, Marconi F, Kaegi R, Odzak N, Behra R. Toxicity of silver nanoparticles to *Chlamydomonas reinhardtii*. Environmental Science and Technology. 2008;**42(23)**:8959–8964. DOI: 10.1021/es801785m

[160] Andoscha A, Affenzeller MJ, Lütz C, Lütz-Meindl U. A freshwater green alga under cadmium stress: ameliorating calcium effects on ultrastructure and photosynthesis in the unicellular model *Micrasterias*. Journal of Plant Physiology. 2012;**169**:1489–1500. DOI: 10.1016/j.jplph.2012.06.002

[161] Machado MD, Lopes AR, Soares EV. Responses of the alga *Pseudokirchneriella subcapitata* to long-term exposure to metal stress. Journal of Hazardous Materials. 2015;**296**:82–92. DOI: 10.1016/j.jhazmat.2015.04.022

[162] Bakar LA, Amina NM, Zakeria HM. Accumulation of Cu(II) and Pb(II) in three rhodophytes of the genus *Gracilaria* and the impact of the metals on the algal physiology. Bioscience and Bioengineering. 2015;**1(4)**:106–111. ISSN: 2381–7704 (Online). Available from: https://www.researchgate.net/publication/301363571_Accumulation_of_CuII_and_PbII_in_Three_Rhodophytes_of_the_Genus_Gracilaria_and_the_Impact_of_the_Metals_on_the_Algal_Physiology [Accessed: 2016-12-20].

[163] Deep PR, Bhattacharyya S, Nayak B. Effect on biochemical parameters of cyanobacterium *Anabaena* sp under lead stress. International Journal of Advanced Research. 2016;**4(9)**:2114–2129. DOI: 10.21474/IJAR01/1697

[164] Deckert J. Cadmium toxicity in plants: is there any analogy to its carcinogenic effect in mammalian cells? Biometals. 2005;**18**:475–481. DOI: 10.1007/s10534-005-1245-0

[165] Faller P, Kienzler K, Liszkay AK. 2005. Mechanism of Cd2+ toxicity: Cd^{2+} inhibits photoactivation of photosystem II by competitive binding to the essential Ca^{2+} site. Biochimica et Biophysica Acta. 2005;**1706**:158–164. DOI: 10.1016/j.bbabio.2004.10.005

[166] Shanker AK, Cervantes C, Loza-Tavera H, Avudainayagam S. Chromium toxicity in plants. Environmental International. 2005;**31(5)**:739–753. DOI: 10.1016/j.envint.2005.02.003

[167] Horcsik ZT, Kovacs L, Laposi R, Meszaros I, Lakatos G, Garab G. Effect of chromium on photosystem 2 in the unicellular green alga, *Chlorella pyrenoidosa*. Photosynthetica. 2007;**45(1)**:65–69. DOI: 10.1007/s11099-007-0010-8

[168] Juarez AB, Barsanti L, Passarelli V, Evangelista V, Vesentini N, Conforti V, Gualtieri P. In vivo microspectroscopy monitoring of chromium effects on the photosynthetic and photoreceptive apparatus of *Eudorina unicocca* and *Chlorella Kessleri*. Journal of Environmental Monitoring. 2008;**10 (11)**:1313–1318. DOI: 10.1039/b809566c

[169] Zemri K, Amar Y, Boutiba Z, Zemri M, Popovic R. Use of chlorophyll fluorescence to evaluate the effect of chromium on activity photosystem II at the alga *Scenedesmus obliquus*. International Journal of Research and Reviews in Applied Science. 2012;**12(2)**:304–314. Available from: http://www.arpapress.com/Volumes/Vol12Issue2/IJRRAS_12_2_18.pdf [Accessed: 2017-01-02]

[170] Takami R, Almeida JV, Vardaris CV, Colepicolo P, Barrosa MP. The interplay between thiol-compounds against chromium (VI) in the freshwater green alga *Monoraphidium convolutum*: toxicology, photosynthesis, and oxidative stress at a glance. Aquatic Toxicology.2012;**118–119**:80–87. DOI: 10.1016/j.aquatox.2012.03.018

[171] Wong PK, Chang L. Effects of copper, chromium and nickel on growth, photosynthesis and chlorophyll a synthesis of *Chlorella pyrenoidosa* 251. Environmental Pollution. 1991;**72(2)**:127–139. DOI: 10.1016/0269-7491(91)90063-3

[172] Robinson NJ. Algal metallothioneins: secondary metabolites and proteins. Journal of Applied Phycology. 1989;**1(1)**:5–18. DOI: 10.1007/BF00003530

[173] Nordberg M, Nordberg GF. Metallothioneins: historical development and overview. In: Sigel A, Sigel H, Sigel RKO, editors. Metal Ions in Life Sciences, Vol. 5: Metallothioneins and Related Chelators. Cambridge, England: Royal Society of Chemistry. 2009. pp. 1–29. ISBN:1-84755-899-2

[174] Ajayan KV, Selvaraju M. Heavy metal induced antioxidant defense system of green microalgae and its effective role in phycoremediation of tannery effluent. Pakistan Journal of Biological Sciences. 2012;**15**:1056–1062. DOI: 10.3923/pjbs.2012.1056.1062

[175] Vagi MC, Kostopoulou MN, Petsas AS, Lalousi ME, Rasouli C, Lekkas TD. Toxicity of organophoshorous pesticides to the marine alga *Tetraselmis suecica*. Global NEST Journal. 2005;**7(2)**:222–227. Available from: http://journal.gnest.org/sites/default/files/Journal%20Papers/paper_9_VAGI_375.pdf [Accessed: 2017-01-23]

[176] Rimet F, Bouchez A. Use of diatom life-forms and ecological guilds to assess pesticide contamination in rivers: lotic mesocosm approaches. Ecological Indicators. 2011;**11**:489–499. DOI: 10.1016/j.ecolind.2010.07.004

Chapter 5

Effects of pH and Phosphorus Concentrations on the Chlorophyll Responses of *Salvia chamelaeagnea* (Lamiaceae) Grown in Hydroponics

Kerwin Lefever, Charles P. Laubscher,
Patrick A. Ndakidemi and Felix Nchu

Additional information is available at the end of the chapter

http://dx.doi.org/10.5772/67610

Abstract

Salvia chamelaeagnea (Lamiaceae) is a slow growing water-wise evergreen shrub originating from the western province of South Africa. It is an attractive landscape, and *S. chamelaeagnea* is a medicinal plant. It is important to develop enhanced cultivation protocols that could result in high yield and high-quality medicinal materials. Chlorophyll is a fundamental part of the light-dependent reactions of the photosynthesis process. This chapter investigates the effects of four phosphorus concentrations and three pH levels of supplied irrigated water on the production of chlorophyll A, chlorophyll B, total chlorophyll, leaf colour and the nutrient uptake of *S. chamelaeagnea* grown in hydroponics over an 8-week period at the Cape Peninsula University of Technology. The treatments of pH 4, pH 6 and pH 8 at 31, 90, 150 and 210 ppm of phosphorus were received by 12 groups of plants and were replicated 10 times. The results indicated that at pH 4, P fertilization significantly ($P < 0.05$) induced a higher chlorophyll production of *S. chamelaeagnea* grown in hydroponics compared to other pH treatments (pH 8 and pH 6).

Keywords: hydroponics, pH, chlorophyll production, medicinal plants, *Salvia chamelaeagnea*

1. Introduction

Salvia chamelaeagnea P.J. is a member of the Lamiaceae family. Plant species in this family include many culinary and medicinal herbs like *Salvia officinalis, Salvia verbenacea* and *Salvia libanotica,* which have been used for many years against diarrhoea, indigestion, colic, abdominal trouble, influenza, bacterial infections, tuberculosis, cough, cold and many other ailments [1–3]. Some of these uses date back to medieval times [4]. Many of the Lamiaceae secondary metabolites are of commercial interest to the food industry as sources of natural preservatives, flavourants and antioxidants [2, 5], as well as to the pharmaceutical industry as sources of antioxidants, anti-inflammatories [6], antibacterials and anti-mycobacterials [7].

Salvias are renowned for their variety and their many uses around the home and garden; they have beautiful flowers and attract birds [8]. In its natural habitat, *S. chamelaeagnea* will develop into attractive foliage and flowering landscape plants, with small mid-green egg-shaped leaves and masses of bright blue or white flowers borne at the tops of each stem, which are suitable for the cut flower trade [8–10]. *S. chamelaeagnea* also has value in the medicinal plant trade as it contains the phenolic compounds carnosol, rosmarinic acid and caffeic acid, which exhibits antioxidant and anti-bacterial activities [2, 6, 11].

Unfortunately, very little information has been documented on the cultivation of this species. Cultivation of medicinal plants is gaining traction worldwide; it is seen as a tool for biodiversity conservation, poverty alleviation and cultural preservation [12]. However, good knowledge of plant physiology must be attained in order to develop enhanced cultivation protocols that could result in high yield and high-quality medicinal materials. Effects of nutrients and nutrient ratios on many food and medicinal crop plants, such as soya bean, thyme, wheat cultivars, barley, spinach and pelargoniums, have been studied. In most cases, a positive result in growth is noticed with the addition of some macro-nutrients such as N, P, K, Mg or Ca [13–21]. It is therefore crucial that adequate plant nutrition and soil pH levels are met for any given plant so that the cell's functioning is not impeded. Chlorophyll is a fundamental part of the light-dependent reactions of the photosynthesis process, capturing light rays from the sun and producing energy-storing ATP molecules that are essential for the functioning of a healthy plant [22, 23]. The effects of poor nutrition, be it through infertile soils or incorrect soil pH level, directly affect the production of chlorophyll molecules resulting in chlorosis of leaves and a reduced photosynthetic rate, thus inhibiting some biological processes and decreasing the general health of the plants [23–25]. There are plausible mechanisms through which the production of chlorophyll could be affected, for example, the pH level of a growing medium affects the uptake of P [26] and the P level influences the nutrient uptake by plants [27]. The relationship between the nutrient P and chlorophyll is not fully understood. According to Nicholls and Dillon [28], there are substantial variations of the published phosphorus-chlorophyll relationship, which they ascribed to variations in sampling and analytical techniques.

This chapter aims to investigate the effects of P and pH on the chlorophyll production, leaf colour and the nutrient uptake of medicinal *S. chamelaeagnea* in hydroponics, in order to determine a fertilizer regime that will promote the development of *S. chamelaeagnea* without degrading soils and leaching nutrients into the water table.

2. Materials and methods

2.1. Experimental process

The experiment took place in the research glasshouse at the Cape Peninsula University of Technology (CPUT), Cape Town campus, South Africa, latitude and longitude S33°55′ 58

http://dx.doi.org/10.5772/67610

E18°25′ 57, from June 2012 to August 2012. Inside the glasshouse was a 40%-Aluminet shade cloth, raised 2 m above the floor, resulting in light intensities ranging from 10 to 13 Klx, determined by using a Toptronic T630 light meter. The climate was controlled between 16 and 28°C during the day while 10–20°C during the night, with an average relative humidity of 42%.

The experiment was laid out in a randomized block design with plants being spaced 30 cm apart and consisted of 12 treatments of four differing nutrient solutions offering a low concentration of P, a balanced concentration of supplementary P, a moderate concentration of supplementary P and a high concentration of supplementary P at three differing pH levels. The control treatment of 31 ppm was chosen due to the nature of fynbos soils being low in available P [29–31].

Hoagland solution, a well-known hydroponic nutrient solution modified by Hershey [32, 33], offering all the necessary macro- and micro-nutrients for healthy plant growth, was used as a base nutrient and supplemented with P.

The plants for the experiment were rooted tip cuttings sourced from healthy mother stock plants at the CPUT Glass House Nursery. The rooted cuttings were gently rinsed in deionized water to remove any rooting media from the root's zone. They were then weighed and planted into 25-cm plastic pots filled with leca clay and placed into a recirculating closed hydroponics system at a spacing of 30 cm, where their heights were recorded (**Figure 1**).

Figure 1. *S. chamelaeagnea* rooted cuttings exposed to varied combinations of pH and P treatments in hydroponics under greenhouse conditions (Picture: K. Lefever).

The plants were irrigated with the treatments 15 times per day at equal timed intervals for the duration of the experiment. For each treatment, there were 10 plants. The treatments were as follows:

1. Hoagland hydroponic nutrient solution with 31 ppm of P at a pH of 4.
   - Hoagland hydroponic nutrient solution with 31 ppm of P at a pH of 6.
   - Hoagland hydroponic nutrient solution with 31 ppm of P at a pH of 8.

2. Hoagland hydroponic nutrient solution supplemented with 90 ppm of P at a pH of 4.
   - Hoagland hydroponic nutrient solution supplemented with 90 ppm of P at a pH of 6.
   - Hoagland hydroponic nutrient solution supplemented with 90 ppm of P at a pH of 8.

3. Hoagland hydroponic nutrient solution supplemented with 150 ppm of P at a pH of 4.
   - Hoagland hydroponic nutrient solution supplemented with 150 ppm of P at a pH of 6.
   - Hoagland hydroponic nutrient solution supplemented with 150 ppm of P at a pH of 8.

4. Hoagland hydroponic nutrient solution supplemented with 210 ppm of P at a pH of 4.
   - Hoagland hydroponic nutrient solution supplemented with 210 ppm of P at a pH of 6.
   - Hoagland hydroponic nutrient solution supplemented with 210 ppm of P at a pH of 8.

2.2. pH level

The pH levels of the nutrient solutions were monitored using a Martini Instrument PH55 pH probe and were adjusted accordingly using either hydrochloric acid (HCl) to lower the pH or sodium hydroxide (NaOH) to raise the pH.

2.3. Irrigation

The treatments were set to irrigate 15 times daily for a duration of 15 min using a 1350 L/h Boyu submersible pump and a Tedelex analogue timer to regulate irrigation frequencies.

2.4. Data collection

2.4.1. Measurement of leaf colour

Green leaf colour intensity was measured using a hand-held, dual-wavelength SPAD meter (SPAD 502, chlorophyll meter, Minolta Camera Co., Ltd., Japan). Readings were taken from the top three fully developed leaves of each plant. For each treatment, 30 fully developed leaves were used weekly. The SPAD meter stored and automatically averaged the recordings to generate one reading per plant.

http://dx.doi.org/10.5772/67610

2.4.2. Measurement of chlorophyll content in leaves

The extraction of leaf chlorophyll using dimethylsulphoxide (DMSO) was carried out as described in Hiscox and Israelsta [34]. A third of plant leaves from the tip were collected from each plant. About 100 mg of the middle portion of the fresh leaf slices was placed in a 15-mL vial containing 7 mL DMSO and incubated at 4°C for 72 h. After the incubation, the extract was diluted to 10 mL with DMSO. A 3-mL sample of chlorophyll extract was then transferred into curvets for absorbance determination. A spectrophotometer (UV/Visible Spectrophotometer, Pharmacia LKB. Ultrospec II E) was used to determine absorbance values at 645 and 663 nm, which were then used in the equation proposed by Arnon [35] to determine the total leaf chlorophyll content against DMSO blank, expressed as mg L^{-1} as follows: Chl a = 12.7D663 - 2.69D645, Chl b = 22.9D645 - 4.68D663 and Total Chl = 20.2D645 + 8.02D663.

2.4.3. Measurement of the levels of macro- and micro-nutrients in dry plant material

The measurements of macronutrients (N, P, K, Ca and Mg) and micronutrients (Cu, Zn, Mn, Fe and B) were determined by ashing a 1 g ground sample in a porcelain crucible at 500°C overnight. This was followed by dissolving the ash in 5 mL of 6 M HCl and putting it in an oven at 50°C for 30 min; 35 mL of deionized water was added, and the extract was filtered through Whatman no. 1 filter paper. Nutrient concentrations in plant extracts were determined using an inductively coupled plasma (ICP) emission spectrophotometer (IRIS/AP HR DUO Thermo Electron Corporation, Franklin, Massachusetts, USA) [36].

2.5. Statistical analysis

Data collected was analysed for statistical significance using the two-way analysis of variance (ANOVA), with the computations being done using the software program STATISTICA. Fisher's least significance difference (LSD) was used to compare treatment means at $P \leq 0.05$ level of significance [37].

3. Results

3.1. Effects of pH and phosphorus concentrations on the chlorophyll content of *S. chamelaeagnea* grown in hydroponics

Treatment significantly ($P \leq 0.001$) affected the chlorophyll A, chlorophyll B and total chlorophyll contents of *S. chamelaeagnea* grown hydroponically (**Table 1**). The chlorophyll A (10.9–12.2), chlorophyll B (3–3.4) and total chlorophyll (13.9–14.7) values of the plants exposed to phosphorus at pH 4 treatments were significantly ($P \leq 0.001$) higher compared to the corresponding values at pH 6 (chlorophyll A [8.3–10.3], chlorophyll B [2.2–2.8] and total chlorophyll [10.7–13.4] and at pH 8—chlorophyll A [3.5–10.17], chlorophyll B [0.91–2.7] and total chlorophyll [4.4–12.9]) treatments (**Table 1**). Leaf chlorosis of plants grown at pH 8 was observed.

Treatments	Chlorophyll A	Chlorophyll B	Total chlorophyll
pH 4, P 31 ppm	12.242 ± 1.7a	3.446 ± 0.5a	15.684 ± 2.2a
pH 6, P 31 ppm (Control)	10.384 ± 1.0cd	2.848 ± 0.3cde	13.229 ± 1.3cd
pH 8, P 31 ppm	10.173 ± 1.1cde	2.784 ± 0.3ef	12.954 ± 1.5cde
pH 4, P 90 ppm	11.419 ± 0.5ab	3.233 ± 0.2ab	14.649 ± 0.6ab
pH 6, P 90 ppm	8.348 ± 1.1g	2.227 ± 0.3hi	10.574 ± 1.4g
pH 8, P 90 ppm	9.327 ± 1.3ef	2.600 ± 0.4g	11.924 ± 1.7ef
pH 4, P 150 ppm	10.929 ± 0.7bc	3.014 ± 0.3bcd	13.941 ± 0.9bc
pH 6, P 150 ppm	8.463 ± 1.4fg	2.282 ± 0.4h	10.744 ± 1.8fg
pH 8, P 150 ppm	7.063 ± 0.6h	1.988 ± 0.2i	9.049 ± 0.7h
pH 4, P 210 ppm	10.900 ± 0.7bc	3.108 ± 0.3bc	14.005 ± 0.9bc
pH 6, P 210 ppm	9.817 ± 1.0de	2.650 ± 0.3g	12.465 ± 1.3de
pH 8, P 210 ppm	3.547 ± 0.5i	0.910 ± 0.2j	4.456 ± 0.7i
One-way ANOVA (*F*-statistic)	46.757***	43.425***	46.388***

[a-j]Means followed by same lowercase letters in the same column are not significantly different ($P > 0.05$) following comparison using Tukey test.

***represents a statistical significance of ($P \leq 0.001$) according to Fisher's least significant difference.

Table 1. The effects of pH and Phosphorus concentrations on the chlorophyll content of *S. chamelaeagnea* grown in hydroponics.

3.2. Effects of pH and phosphorus concentrations on the leaf colour of *S. chamelaeagnea* grown in hydroponics

Effects of various P treatments at differed pH levels induced varied colour intensities, ranging from 16 to 31.7 from week 1 to week 8 on the leaf colour of *S. chamelaeagnea* ($P \leq 0.001$) (**Table 2**, **Figure 2**). While treatment 1 offering a pH level of 4 at 31 ppm P generally yielded the highest leaf colour values over the 8-week growth period, these values did not differ significantly from that of the other pH 4 treatments receiving supplementary P. Of these treatments receiving supplementary P, the highest results were recorded at pH 4 receiving 210 ppm P closely followed by pH 4 at 90 ppm P and pH 4 at 150 ppm P treatments, respectively.

3.3. Effects of pH and phosphorus concentrations on the uptake of macro-nutrients in *S. chamelaeagnea* grown in hydroponics

Macro-nutrient uptake of P, K and Mg was significantly ($P \leq 0.001$) affected by the treatment (**Table 3**). There was a noticeable higher tissue P content (1.07 ± 0.08%) at pH 8, 150 ppm of P (**Table 3**). Tissue nitrogen content (4.41 ± 0.20%) was significantly higher in plants in treatment (90 ppm of P) at pH 6. Highest uptake of Ca was recorded at a pH of 8 at 90 ppm of P.

Treatments	Wk1	Wk2	Wk3	Wk4	Wk5	Wk6	Wk7	Wk8
pH 4, P 31 ppm	30.156 ± 3.6ab	31.667 ± 4.1a	31.433 ± 3.0a	30.878 ± 1.3a	30.189 ± 3.1ab	30.011 ± 2.2a	31.078 ± 1.8a	28.467 ± 1.9ab
pH 6, P 31 ppm (Control)	32.800 ± 2.5a	31.644 ± 4.0a	30.533 ± 1.3ab	29.933 ± 2.0a	30.122 ± 2.4ab	28.822 ± 2.6ab	28.644 ± 1.6bcd	28.922 ± 1.4a
pH 8, P 31 ppm	30.033 ± 4.5ab	30.567 ± 3.8ab	28.933 ± 1.4b	29.944 ± 2.1a	29.189 ± 1.6ab	30.111 ± 2.4a	29.589 ± 1.7abc	28.278 ± 2.3ab
pH4, P 90 ppm	29.689 ± 3.8ab	29.911 ± 5.3ab	30.867 ± 2.2ab	31.111 ± 1.5a	30.711 ± 2.1ab	29.656 ± 2.4a	30.567 ± 2.1a	28.122 ± 2.0ab
pH 6, P 90 ppm	31.789 ± 3.6ab	31.156 ± 4.6ab	31.078 ± 2.8ab	30.578 ± 2.0a	29.444 ± 1.6ab	27.000 ± 1.4b	28.067 ± 2.2cde	27.056 ± 1.0bc
pH 8, P 90 ppm	29.411 ± 3.3b	27.356 ± 3.0bc	22.756 ± 3.0c	20.067 ± 2.3b	20.278 ± 1.2c	24.489 ± 1.6c	27.000 ± 1.9e	27.800 ± 1.7abc
pH 4, P 150 ppm	30.289 ± 3.6ab	30.044 ± 3.7ab	31.233 ± 2.7ab	30.622 ± 1.5a	29.189 ± 2.6ab	29.989 ± 4.0a	30.178 ± 1.8ab	27.956 ± 1.1ab
pH 6, P 150 ppm	29.333 ± 3.5b	30.944 ± 3.6ab	29.933 ± 2.1ab	29.356 ± 2.7a	28.633 ± 2.3b	28.233 ± 2.2ab	26.456 ± 1.5e	25.100 ± 2.1de
pH 8, P 150 ppm	29.267 ± 3.0b	22.422 ± 4.9d	15.978 ± 2.8d	15.756 ± 3.5c	15.156 ± 2.7d	15.267 ± 3.1e	21.756 ± 1.6f	24.278 ± 2.3e
pH 4, P 210 ppm	30.944 ± 2.8ab	30.456 ± 3.1ab	28.956 ± 2.4b	31.033 ± 2.0a	30.911 ± 3.2a	28.767 ± 2.0ab	29.533 ± 0.7abc	29.278 ± 1.4a
pH 6, P 210 ppm	31.756 ± 3.9ab	28.489 ± 3.2abc	29.478 ± 2.4ab	29.444 ± 1.2a	29.722 ± 1.0ab	28.622 ± 1.7ab	27.756 ± 2.0de	26.278 ± 1.7cd
pH 8, P 210 ppm	31.411 ± 2.6ab	24.922 ± 5.7cd	16.167 ± 3.6d	13.900 ± 2.7c	14.511 ± 1.8d	18.044 ± 1.7d	18.567 ± 1.6g	16.067 ± 1.8f
Two-way ANOVA (*F*-statistic)	1.013NS	4.333***	45.53***	78.77***	66.25***	38.79***	41.52***	37.33***

NS represents no statistical significance,
*represents a statistical significance of ($P \leq 0.05$),
**represents a statistical significance of ($P \leq 0.01$) and
***represents a statistical significance of ($P \leq 0.001$) according to Fisher's least significant difference.

Table 2. The effects of pH and phosphorus concentrations on the leaf colour of *S. chamelaeagnea* grown in hydroponics.

3.4. Effects of pH and phosphorus concentrations on the uptake of micro-nutrients in *S. chamelaeagnea* grown in hydroponics

The micro-nutrient uptake of Na, Mn, Fe, Cu, Zn and B was significantly ($P \leq 0.001$) affected by the treatments (**Table 4**). The highest nutrient uptake values of Na (867.67 ± 131.72%) and Zn (46.78 ± 7.31%) were recorded at pH 8, 210 ppm of P treatment. The Fe uptake value (175.00 ± 14.42%) in the treatment at pH 4 of 210 ppm was the highest value. Highest recorded uptake values of Cu were obtained in plants receiving a pH of 4 at 31 ppm of P closely followed by the plants receiving a pH of 4 at 210 ppm.

Figure 2. Observable variations in the leaf's green colour among plants (*S. chamelaeagnea*) following exposure to varied combinations of pH and P treatments in hydroponics under greenhouse conditions (Picture: K. Lefever).

Treatments	N (%)	P (%)	K (%)	Ca (%)	Mg (%)
pH 4, P 31 ppm	4.18 ± 0.29bcd	0.64 ± 0.06g	4.23 ± 0.29g	1.13 ± 0.07a	0.28 ± 0.01h
pH 6, P 31 ppm	4.24 ± 0.55abc	0.73 ± 0.06f	4.41 ± 0.23fg	1.12 ± 0.10a	0.36 ± 0.03e
pH 8, P 31 ppm	4.19 ± 0.18abcd	0.62 ± 0.08g	4.47 ± 0.26efg	1.10 ± 0.11ab	0.43 ± 0.04c
pH 4, P 90 ppm	4.27 ± 0.26abc	0.77 ± 0.08ef	4.64 ± 0.36cdef	1.10 ± 0.10ab	0.31 ± 0.02g
pH 6, P 90 ppm	4.41 ± 0.20a	0.82 ± 0.08cde	4.53 ± 0.13defg	1.08 ± 0.07ab	0.38 ± 0.03d
pH 8, P 90 ppm	4.20 ± 0.12abcd	0.80 ± 0.07def	4.45 ± 0.24efg	1.14 ± 0.05a	0.48 ± 0.03b
pH 4, P 150 ppm	4.37 ± 0.19ab	0.82 ± 0.04cde	4.79 ± 0.58cd	1.07 ± 0.06abc	0.32 ± 0.03fg
pH 6, P 150 ppm	4.09 ± 0.22cd	0.88 ± 0.07bc	4.87 ± 0.19c	1.01 ± 0.07cd	0.36 ± 0.02de
pH 8, P 150 ppm	4.00 ± 0.08de	1.07 ± 0.08a	6.29 ± 0.39a	0.77 ± 0.03e	0.55 ± 0.02a
pH 4, P 210 ppm	4.13 ± 0.18cd	0.84 ± 0.06bcd	4.73 ± 0.32cde	1.05 ± 0.06bc	0.31 ± 0.02g
pH 6, P 210 ppm	4.05 ± 0.18cd	0.87 ± 0.08bcd	4.23 ± 0.35g	0.97 ± 0.06d	0.35 ± 0.02ef
pH 8, P 210 ppm	3.77 ± 0.16e	0.91 ± 0.13b	5.71 ± 0.38b	0.46 ± 0.03f	0.48 ± 0.03b
One-way ANOVA (*F*-statistic)	4.35***	21.34***	31.67***	68.64***	89.74***

NS represents no statistical significance,
*represents a statistical significance of ($P \leq 0.05$),
**represents a statistical significance of ($P \leq 0.01$) and
***represents a statistical significance of ($P \leq 0.001$) according to Fisher's least significant difference.

Table 3. The effects of pH and Phosphorus concentrations on the uptake of macro-nutrients in *S. chamelaeagnea* grown in hydroponics.

Treatments	Na (mg/kg)	Mn (mg/kg)	Fe (mg/kg)	Cu (mg/kg)	Zn (mg/kg)	B (mg/kg)
pH 4, P 31 ppm	477.89 ± 36.27fg	84.67 ± 7.48efg	151.56 ± 7.32cde	5.22 ± 1.99a	39.56 ± 2.88bc	37.78 ± 3.63ab
pH 6, P 31 ppm (Control)	479.78 ± 57.99fg	105.89 ± 11.40c	139.11 ± 10.17def	2.89 ± 0.60d	38.00 ± 3.20c	38.56 ± 3.09a
pH 8, P 31 ppm	472.89 ± 58.58g	156.78 ± 9.11a	137.11 ± 8.25ef	2.89 ± 0.33d	37.89 ± 3.44c	37.33 ± 2.29ab
pH4, P 90 ppm	548.44 ± 74.72ef	84.00 ± 8.19fg	144.11 ± 10.59def	4.11 ± 0.60bc	40.11 ± 4.31bc	38.67 ± 3.67a
pH 6, P 90 ppm	505.78 ± 39.02fg	101.00 ± 4.69cd	153.33 ± 13.87bcd	2.56 ± 0.53de	41.33 ± 6.12bc	37.44 ± 2.40ab
pH 8, P 90 ppm	532.56 ± 70.06efg	150.33 ± 12.56a	167.33 ± 13.27abc	3.22 ± 0.44cd	39.78 ± 6.28bc	36.56 ± 1.24abc
pH 4, P 150 ppm	604.22 ± 102.07de	82.67 ± 9.84g	168.56 ± 23.51ab	4.67 ± 1ab	41.33 ± 6.24bc	37.67 ± 2.29ab
pH 6, P 150 ppm	680.33 ± 55.08bc	94.00 ± 19.68de	151.00 ± 34.86de	3.11 ± 1.90d	38.00 ± 6.75c	35.22 ± 3.03bcd
pH 8, P 150 ppm	716.00 ± 117.06b	131.44 ± 7.32b	152.78 ± 16.20bcde	4.33 ± 0.5ab	39.44 ± 4.48bc	31.67 ± 3.35e
pH 4, P 210 ppm	696.78 ± 69.99bc	82.11 ± 4.43g	175.00 ± 14.42a,	5.11 ± 0.60a	44.00 ± 2.92ab	34.56 ± 2.51cd
pH 6, P 210 ppm	640.89 ± 36.55cd	90.78 ± 9.38efg	145.11 ± 14.16def	2.44 ± 0.53de	43.89 ± 5.69ab	35.33 ± 2.45bcd
pH 8, P 210 ppm	867.67 ± 131.72a	93.44 ± 8.14def	129.33 ± 21.17f	1.89 ± 0.60e	46.78 ± 7.31a	33.33 ± 2.40de
One-way ANOVA (*F*-statistic)	22.746***	62.30***	5.590***	11.975***	2.573***	5.56***

NS represents no statistical significance,
*represents a statistical significance of ($P \leq 0.05$),
**represents a statistical significance of ($P \leq 0.01$) and
***represents a statistical significance of ($P \leq 0.001$) according to Fisher's least significant difference.

Table 4. The effects of pH and phosphorus concentrations on the uptake of micro-nutrients in *S. chamelaeagnea* grown in hydroponics.

4. Discussions

In this chapter, the significantly ($P \leq 0.001$) higher chlorophyll values recorded in the treatments at a pH of 4 with supplementary P show that phosphorous fertilization under an acidic condition of chlorophyll production by *S. chamelaeagnea* will largely increase in hydroponic production. Also, high leaf colour intensity values were recorded in treatments with a pH of 4 compared to that of treatments with a higher pH of 6 or 8. On the other hand, it seems that a higher P concentration had a minimal effect on leaf colour intensity. It is worth noting that studies have shown high correlations between chlorophyll meter readings, that is, the leaf's green colour intensity and extractable leaf chlorophyll [38]. The effect of P on chlorophyll could be indirect and complex. P fertilization may indirectly influence or hinder the uptake of other nutrients [39], which in turn affects chlorophyll production in plants. Indigenous plants, especially those occurring in the fynbos biome, are expected to be adapted to nutrient-poor and low-pH soils and tend to have low critical levels for most of the nutrients. Therefore, exposing these species to high P concentration may have a minimal effect on plant physiology and can even have detrimental effects on plant growth.

Despite the relatively high nutrient uptake values in plants receiving a nutrient solution with a pH 8, chlorosis of their leaves was apparent during the growth period. This suggests that the uptake of some essential nutrients responsible for chlorophyll development was affected at this pH level, namely the mineral nutrients Cu, B, N and Fe which are directly involved in photosynthesis, respiration, cell division and protein formation [23, 40]. In soil-less media, the affinity of soluble nutrients to negatively charged surfaces and the interactions between charged cations can have a profound effect on nutrient availability and subsequently, the uptake of nutrients by plants. For example, fertilization with phosphorous increases the soil's nitrogen absorption in young plants of *Eucalyptus grandis* [39]. Silber [41] argued that a continuous decline of soluble P concentration during fertilization can be explained through two mechanisms, a rapid electrostatic reaction and adsorption of the onto substrate and a slow formation of solid metal-P compounds with Al and Fe under acidic conditions and Ca and Mg under basic-to-neutral conditions. Therefore, the substrate used in hydroponic setups could affect the availability of micro- and macro-nutrients. Shen et al. [42] suggested that the availability of soil P is extremely complex and needs to be systemically evaluated. Previously, Wu et al. [43] showed that under phosphorus stress, no significant changes in chlorophyll A and B, total chlorophyll and carotenoid contents were found, and phosphorus stress generally had no effect on photosynthesis. The highest nutrient uptake values were recorded in nine of the 12 treatments receiving supplementary P, with only Cu and Mn yielding the highest values in treatments receiving no supplementary P. Thus, it is evident that phosphorus treatments had a significant effect on nutrient uptake in *S. chamelaeagnea* grown hydroponically [44].

5. Conclusions

In conclusion, this chapter gives insight into the unknown cultivation requirements of the leaf's chlorophyll development of *S. chamelaeagnea* and shows that the use of a hydroponic nutrient system offering little to no supplementary phosphorus at a pH level of 4 significantly correlated with the chlorophyll development of *S. chamelaeagnea* grown in hydroponics. Based on the results obtained in the chapter, it is plausible to assume that P has an indirect effect on chlorophyll production in *S. chamelaeagnea*.

Acknowledgements

This study was funded by Cape Peninsula University of Technology through CPUT Bursary and University Research Funds.

Author details

Kerwin Lefever[1], Charles P. Laubscher[1]*, Patrick A. Ndakidemi[2] and Felix Nchu[1]

*Address all correspondence to: laubscherc@cput.ac.za

1 Faculty of Applied Sciences, Cape Peninsula University of Technology, Bellville, South Africa

2 School of Life Sciences and Bio-engineering, Nelson Mandela Institute of Science and Technology, Arusha, Tanzania

References

[1] Gali-Muhtasib, H., Hilan, C. & Khater, C. 2000. Traditional uses of *Salvia libanotica* (East Mediterranean sage) and the effects of its essential oils. *Journal of Ethnopharmacology*. 71(3):513–520.

[2] Kamatou, G.P.P., van Zyl, R.L., Davids, H., van Heerden, F.R., Lourens, A.C.U. & Viljoen. A.M. 2008. Antimalarial and anticancer activities of selected South African Salvia species and isolated compounds from *S. radula*. *South African Journal of Botany*. 74:238–243.

[3] Cornara, L., La Rocca, A., Marsili, S. & Mariotti, M.G. 2009. Traditional uses of plants in the Eastern Riviera (Liguria, Italy). *Journal of Ethnopharmacology*. 125(1):16–30.

[4] Clebsch, B. 2003. The new book of Salvias, 2nd ed. Port Land: Timber Press Inc., pp. 1–80.

[5] Thorsen, M.A. & Hildebrandt, K.S. 2003. Quantitative determination of phenolic diterpenes in rosemary extracts, aspects of accurate quantification. *Journal of Chromatography*. 995:119–125.

[6] Kamatou, G.P.P., Viljoen, A.M. & Steenkamp, P. 2010. Antioxidant, anti-inflammatory activities and HPLC analysis of South African Salvia species. *Food Chemistry*. 119(2):684–688.

[7] Kamatou, G.P.P., van Vuuren, S.F., van Heerden, F.R., Seaman, T. & Viljoen A.M. 2007. Antibacterial and antimycobacterial activities of South African Salvia species and isolated compounds from *S. chamelaeagnea*. *South African Journal of Botany*. 73(4):552–557.

[8] Western, P. & Claßen-Bockhoff, R. 2006. Bird pollination in South African Salvia species. *Flora*. 201:396–406.

[9] Pienaar, K. 2000. What flower is that? 2nd ed. Cape Town: Struik, pp. 297–298.

[10] Ulubelen, A. 2003. Cardio active and antibacterial terpenoids from some Salvia species. *Phytochemistry*. 64:395–399.

[11] Nenadis, N., Zafiropoulou, I. & Tsimidou, M. 2003. Commonly used food antioxidants: a comparative study in dispersed systems. *Food Chemistry*. 82:403–407.

[12] Wiersum, K.F., Dold, A.P., Husselman, M. & Cocks, M. 2006. Cultivation of medicinal plants as a tool for biodiversity conservation and poverty alleviation in the amatola

region, South Africa. In R.J. Bogers (ed.), *Medicinal and aromatic plants*. The Netherlands: Springer, pp. xx–xxx. http://bioculturaldiversity.co.za/publications/CULTIVATION%20 OF%20MEDICINAL%20PLANTS%20AS%20A%20TOOL.pdf

[13] Gan, Y., Stulen, I., van Keulen, H. & Kuiper, P.J.C. 2003. Effect of N fertilizer top-dressing at various reproductive stages on growth, N_2 fixation and yield of three Soybean (*Glycine max* (L.) Merr.) genotypes. *Field Crops Research*. 80:147–155.

[14] Khazaie, H.R., Nadjafi, F. & Bannayan, M. 2008. Effect of irrigation frequency and planting density on herbage biomass and oil production of thyme (*Thymus vulgaris*) and hyssop (*Hyssopus officinalis*). *Industrial Crops and Products*. 27:315–321.

[15] Ichir, L.L., Ismaili, M. & van Cleemput, O. 2003. Effect of organic and mineral fertilizers on N-use by wheat under different irrigation frequencies. *Competes Rendus Biologies*. 326:391–399.

[16] Maidl, F.X., Panse, A., Dennert, J., Ruser, R. & Fischbeck, G. 1996. Effect of varied N rates and N timings on yield, N uptake and fertilizer N use efficiency of a six-row and a two-row winter barley. *European Journal of Agronomy*. 5:247–257.

[17] Stagnari, F., Bitetto, V.D. & Pisante, M. 2007. Effects of N fertilizers and rates on yield, safety and nutrients in processing spinach genotypes. *Scientia Horticulturae*. 114:225–233.

[18] Manzanares, M., Tenorio, J.L. & Ayerbe, L. 1997. Sowing time, cultivar, plant population and application of N fertilizer on Kenaf in Spain's central plateau. *Biomass and Bioenergy*. 12(4):263–271.

[19] Liu, X., Ren, G. & Shi, Y. 2011. The effect of organic manure and chemical fertilizer on growth and development of *Stevia rebaudiana* Bertoni. *Energy Procedia*. 5:1200–1204.

[20] Ram, M., Ram, D. & Roy, S.K. 2003. Influence of an organic mulching on fertilizer nitrogen use efficiency and herb and essential oil yields in geranium (*Pelargonium graveolens*). *Bioresource Technology*. 87:273–278.

[21] Zhu, Z., Liang, Z., Han, R. & Dong, J. 2009. Growth and saikosaponin production of the medicinal herb *Bupleurum chinense* DC. Under different levels of nitrogen and phosphorus. *Industrial Crops and Products*. 29:96–101.

[22] Glass, A.D.M. 1989. Plant nutrition. An introduction to current concepts. Boston: Jones and Bartlett Publishers, Inc., pp. 30–50, 140–230.

[23] Stern, K. 2006. Introductory plant biology, 10th ed. New York: McGraw-Hill, pp. 81, 193–201.

[24] Leopold, A.C. & Kriedemann, P.E. 1975. Plant growth and development, 2nd ed. USA: McGraw-Hill, pp. 133, 417, 439–442.

[25] Marschner, H. 1995. Mineral nutrition of higher plants. London: Academic Press Limited, pp. 605–609, 641–645.

[26] Hipps, N.A., Davies, M.J., Dodds, P. & Buckley, G.P. The effects of phosphorous nutrition and soil pH on growth of some ancient woodland indicator plants and their interaction with competitor species. *Plant and Soil*. 271:131–141.

[27] Li, Y., Wang, T., Li, J. & Ao, Y. 2010. Effects of phosphorus on celery growth and nutrient uptake under different calcium and magnesium levels in substrate culture. *Horticultural Science*. 37:99–108.

[28] Nicholls, K.H. & Dillon, P.J. 1978. An evaluation of phosphorus-chlorophyll-phytoplankton relationships for lakes. *International Review of Hydrology* 63:141–154.

[29] Hawkins, H.K., Hettasch, H., Mesjasz-Przybylowicz, J., Przybylowicz, W. & Cramer, M.D. 2008. Phosphorus toxicity in the Proteaceae: a problem in post-agricultural lands. *Scientia Horticulturae*. 117:357–365.

[30] Rebelo, A.G. 2001. Proteas. A field guide to the Proteas of Southern Africa. South Africa: Fernwood Press, pp. 180–235.

[31] Witkowski, E.T.F. & Mitchell, D.T. 1987. Variations in soil phosphorus in the Fynbos biome. *South African Journal of Ecology*. 72–76:1100–1200.

[32] Hershey, D.R. 1994. Solution culture hydroponics: history and inexpensive equipment. *American Biology Teacher*. 56:111–118.

[33] Hershey, D.R. 1995. Plant biology science projects. New York: Wiley, pp. 20–41.

[34] Hiscox, J.D. & Israelstam, G.F. 1979. A method for the extraction of chlorophyll from leaf tissue without maceration. *Canadian Journal of Botany*. 57:1332–1334.

[35] Arnon, D.I. 1949. Copper enzymes in chloroplasts. Polyphenoloxidase in *Beta vulgaris*. *Plant Physiology*. 24:1–15.

[36] Giron, H.C. 1973. Comparison between dry ashing and wet digestion in the preparation of plant materials for atomic absorption analysis. *Atomic Absorption Newsletter*. 12:28–29.

[37] Steel, R.G.D. & Torrie, J.H. 1980. Principles and procedures of statistics: a biometrical approach, 2nd edition. New York: McGraw-Hill Inc.

[38] Wood C.W., Reeves, D.W. & Hilmerick, D.G. 1993. Relationships between chlorophyll meter readings and leaf chlorophyll concentration, N status, and crop yield: a review. *Proceedings Agronomy Society of New Zealand*. 23: 1-9

[39] Graciano C., Goya, J.F., Frangi J.L. & Guiamet, J.J. 2006. Fertilization with phosphorus increases soil nitrogen absorption in young plants of *Eucalyptus grandis*. *Forest Ecology and Management*. 236:202–210.

[40] Molassiotis, A., Tanou, G., Diamantidis, G., Patakas, A. & Therios, I. 2006. Effects of 4-month Fe deficiency exposure on Fe reduction mechanism, photosynthetic gas exchange, chlorophyll fluorescence and antioxidant defense in two peach rootstocks differing in Fe deficiency tolerance. *Journal of Plant Physiology*. 163:176–185.

[41] Silber, A. 2008. Chemical characteristic of soilless media. In: Michael Raviv and J. Heinrich Lieth, editors. Soilless culture: theory and practice. San Diego: Academic Press, 2008, pp. 209–244.

[42] Shen, J., Yuan, L., Zhang, L., Li, H., Bai, Z., Chen, X., Zhang, W. & Zhang, F. 2011. Phosphorus dynamics: From soil to plant. *Plant Physiology*. 156:997–1005.

[43] Wu C., Fan Z. & Wang, Z. 2004. Effect of phosphorus stress on chlorophyll biosynthesis, photosynthesis and biomass partitioning pattern of *Fraxinus mandchurica* seedlings]. *The Journal of Applied Ecology*. 15(6):935–940.

[44] Naeem, M., Khan, M.M.A., Idrees, M. & Aftab, T. 2010. Phosphorus ameliorates crop productivity, photosynthetic efficiency, nitrogen-fixation, activities of the enzymes and content of nutraceuticals of *Lablab purpureus* L. *Scientia Horticulturae*. 126:205–214.

Chapter 6

Light-Emitting Diodes: Progress in Plant Micropropagation

Jericó J. Bello-Bello, Juan A. Pérez-Sato,
Carlos A. Cruz-Cruz and Eduardo Martínez-Estrada

Additional information is available at the end of the chapter

http://dx.doi.org/10.5772/67913

Abstract

In commercial micropropagation laboratories, the light source is one of the most important factors controlling plant morphogenesis and metabolism of plant cells and tissue and organ cultures. Lamp manufacturers have begun to rate lamps specifically for plant needs. The traditional light source used for in vitro propagation is fluorescent lamps (FLs). However, power consumption in FL use is expensive and produces a wide range of wavelengths (350–750 nm) unnecessary for plant development. Light-emitting diodes (LEDs) have recently emerged as an alternative for commercial micropropagation. The flexibility of matching LED wavelengths to plant photoreceptors may provide more optimal production, influencing plant morphology and chlorophyll content. Although previous reports have confirmed physiological effects of LED light quality on morphogenesis and growth of several plantlets in vitro, these study results showed that LED light is more suitable for plant morphogenesis and growth than FLs. However, the responses vary according to plant species. This chapter describes the applications and benefits of LED lamps on chlorophyll in plant micropropagation. Two study cases are exposed, Anthurium (*Anthurium andreanum*) and moth orchids (*Phalaenopsisis* sp.), both species with economic importance as ornamental plants, where LEDs have a positive effect on in vitro development and chlorophyll content.

Keywords: in vitro cloning, light quality, tissue culture, chlorophyll

1. Introduction

Micropropagation or in vitro plant cloning is being widely used for large-scale plant multiplication. This method enables the identical reproduction of the selected parents, following the maintenance of genetic fidelity. In commercial micropropagation laboratories, the light source is one of the most important factors controlling plant morphogenesis and growth cells, tissue and organ cultures. Lamp manufacturers have begun to rate lamps specifically for plant needs. The traditional light source used for in vitro propagation is fluorescent lamps (FLs). Nevertheless, the power consumption in FLs is expensive and produces a wide range of wavelengths (350–750 nm) unnecessary for plant development. Light-emitting diodes (LEDs) have recently emerged as an alternative for commercial micropropagation. LEDs possess advantages such as less heat radiation, a monochromatic spectrum, greater durability, and low power consumption. The LED illumination system for in vitro culture provides light in the spectral region that is involved in photosynthesis and in the photomorphogenic responses in plants.

LED colors or combinations commonly used for in vitro culture are white, red, blue, and mixture rates of blue and red. It has been reported that red light is important for shoot and stem elongation, phytochrome responses and changes in plant anatomy [1]. In contrast, blue light is important in chlorophyll biosynthesis, stomatal opening, chloroplast maturation, and photosynthesis [2]. Blue and red combination LEDs have been used for studies in many areas of photobiological research such as photosynthesis [3] and chlorophyll synthesis [4].

In addition, several studies have shown positive effects of LED lamps on plant development during in vitro culture of different species such as *Fragaria × ananassa* [5, 6], *Musa* spp. [7], *Solanum tuberosum* [8], *Chrysanthemum* [9, 10], *Vitis riparia × V. vinifera* [11], *Brassica napus* [12], *Populus euroamericana* [13], and *Saccharum* spp. [14], among others. However, the response in LED systems depends on the wavelength to which the plants are exposed and varies according to the species [15].

This chapter describes the applications and benefits of LED lamps on chlorophyll in plant micropropagation. Two study cases are exposed, Anthurium (*Anthurium andreanum* Lind.) and moth orchids (*Phalaenopsisis* sp.), both species with economic importance as ornamental plants, where LEDs have had a positive effect on in vitro development and chlorophyll content.

2. Plant micropropagation

Micropropagation is the asexual propagation of plants using the techniques of plant tissue culture (PTC). Plant tissue culture refers to growing and differentiation of cells, tissues, and organs isolated from the mother plant, on artificial solid or liquid media under aseptic and controlled conditions. The small organs or pieces of tissue plants used in PTC are called explants. Plant tissue culture medium provides inorganic nutrients and usually a carbohydrate to replace the carbon which the plant normally fixes from the atmosphere by photosynthesis. When carbon is supplied with sucrose and kept in low light conditions, micropropagated plantlets are not fully dependent on their own photosynthesis.

The PTC techniques provide a new approach to plant propagation, being the best way to produce uniform plant germplasm and the regeneration of pathogen-free plants. To date, commercial plant micropropagation has shown great productive potential; it is being used in hundreds of commercial laboratories for the propagation of species of agricultural and forestry importance. Commercial micropropagation of different species of economic importance is shown in **Figure 1**.

The commercial micropropagation process is carried out in the following stages:

Stage 0: Mother plant selection. Donor plants are selected and conditioned to be used to initiate in vitro cultures.

Stage I: In vitro establishing. The choice of the explant and its disinfection is carried out to initiate an aseptic culture.

Stage II: Multiplication. It is at this stage that mass propagation is performed, obtaining a large number of new individuals from minimal amounts of tissue.

Stage III: Elongation and rooting. The shoots must form their root system and at the same time increase their size to facilitate their manipulation and adaptation to the acclimatization conditions.

Stage IV: Acclimatization. It consists of a slow reduction of the relative humidity and gradual increases in the luminous intensity for a better adaptation to the external environment.

Figure 1. Commercial micropropagation of different species. (a) *Stevia rebaudiana*, (b) *Ananas comosus*, (c) *Vanilla planifolia* and (d) *Anthurium andreanum*.

Requirements for the completion of each stage of micropropagation vary according to the method being utilized; it is not always necessary to follow each of the prescribed steps.

However, there are factors that affect the micropropagation process, including:

Factors that depend on the explant: Size, physiological age of the tissue, and explant position.

Factors that depend on the culture medium: Growth regulators, macro- and micronutrients, organic nitrogen, and carbon source.

Factors related to the incubation environment: Photoperiod, temperature, humidity, and light source.

Factors related to the incubation environment refer to incubators or growth rooms where temperature, humidity, and light can be controlled. In commercial micropropagation laboratories, the light source is one of the most important factors controlling plant development. Light quality (spectral quality), quantity, (photon flux) and photoperiod have a profound influence on the morphogenesis, growth and chlorophyll contents of a plant cell, and tissue and organ cultures.

The illumination systems allow wavelengths to be matched to plant photoreceptors to provide more optimal production and to influence plant morphology and metabolic composition [16]. Plants use energy between 400 and 700 nm and light in this region is called photosynthetically active radiation (PAR).

The growth and development of plants is dependent on light for:

Photosynthesis: The process whereby light energy is converted to chemical energy in the biosynthesis of chemicals from carbon dioxide and water.

Photomorphogenesis: The light-induced development of structure or form.

Phototropism: The growth response of plants which is induced by unilateral light.

In recent years, LEDs have emerged as an alternative for commercial micropropagation. LEDs possess various advantages such as less heat radiation, small mass, a monochromatic spectrum, greater durability, low power consumption, and specific wavelength. The flexibility of matching LED wavelengths to plant photoreceptors may provide more optimal production, influencing plant morphology and metabolism.

3. Spectral quality of LEDs

The traditional light source used for in vitro propagation is fluorescent lamps (FLs). However, power consumption in FL use is expensive and produces a wide range of wavelengths (350–750 nm) unnecessary for plant development, whereas monochromatic light-emitting diodes (LEDs) emit light at specific wavelengths. In this sense, LEDs can be fine-tuned to only produce the spectrums that plants need for morphogenic responses [17]. The response to LED light in micropropagation systems depends on light irradiance, photoperiod, and wavelength. The wavelength to which in vitro plants are exposed varies according to the species. Recent studies compare the effect of FLs (545–610 nm) vs white LEDs (460 and 560 nm), red LEDs (660 nm), blue LEDs (460 nm), and the combination of blue and red LED (460 and 660 nm) treatments. LEDs affect in vitro rooting, number and length of new shoots, chlorophyll and carotenoid pigments, and other characteristics in plants. The spectral irradiance of LEDs is shown in **Figure 2**.

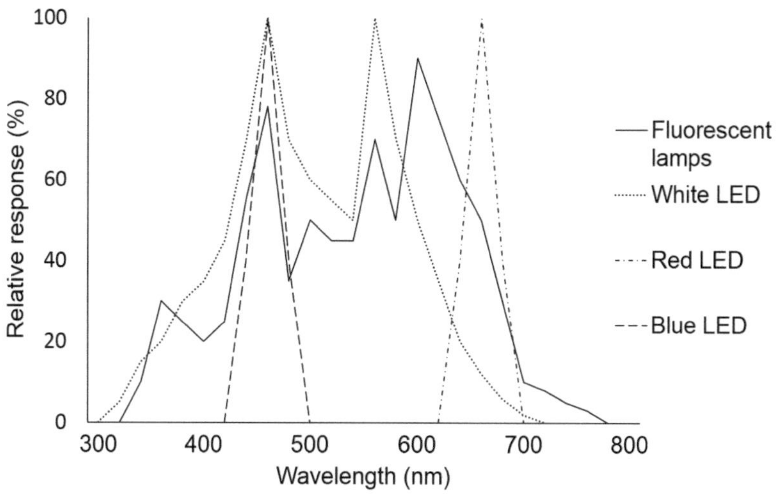

Figure 2. Spectral curves distribution in relative response of the LEDs and fluorescent lamps.

4. LEDs affect chlorophyll content

Several studies have shown important effects of LEDs on photosynthetic pigments during micropropagation of different species. Studies show that blue LEDs are a good light source for chlorophyll induction and that red LEDs decrease chlorophyll content. Dewir et al. [15] found that blue LEDs showed greater growth, vigor, and chlorophyll content in *Euphorbia milli*. Jao et al. [18] reported that blue LEDs promote growth and increase chlorophyll content in *Zantedeschia jucunda*. The same effect was observed by Li et al. [19, 20] during in vitro culture of *Gossypium hirsutum* and *Brassica campestris*, respectively. Kim et al. [9] and Moon et al. [21] emphasized the role of blue light on chlorophyll formation and chloroplast development in their work with *Chrysanthemum* and *Tripterospermum japonicum*, respectively. Monochromatic red LEDs with narrow peak emissions may cause an imbalance in the distribution of light energy between photosystems I and II, and thus be responsible for a reduction in net photosynthesis [3]. According to Li et al. [19], it has been observed that plantlets with lower chlorophyll content utilize the chlorophyll more efficiently than plantlets with higher chlorophyll content under red LEDs.

According to Soebo et al. [22], the possibility exists that red light may inhibit the translocation of photosynthetic products thereby increasing the accumulation of starch. Goins et al. [23]

observed higher photosynthetic rates and an increase in stomatal conductance in wheat leaves under mixed red and blue LEDs. Plant growth and development by increasing net photosynthetic rate was also observed in *Chrysanthemum* under mixed red/blue LED treatments and has been attributed to the similarities of the spectral energy distribution of red/blue to chlorophyll absorption [9].

The importance of blue light in stomatal opening has already been studied. It has been proposed that blue light received by phototropins activates a signaling cascade, resulting in fast stomata opening under a red light background [19]. The effect of light quality on stomatal characteristics has not yet been clearly determined, and differential stomatal behavior could be related to photosynthetic activity and plant growth.

According to Topchiy et al. [24], light quality also plays an important role in photosynthesis, influencing the way in which light is absorbed by chlorophyll. According to George [25], the level of chlorophyll so far obtained in tissue cultures is well below that found in mesophyll cells of whole plants of the same species, and the rate of chlorophyll formation on exposure of cultured cells to the light is extremely slow compared to the response of etiolated organized tissues. The greening of cultures also tends to be unpredictable, and even within individual cells, a range in the degree of chloroplast development is often found. In the carbon dioxide concentrations found in culture vessels, green callus tissue is normally photomixotrophic and growth is still partly dependent on the incorporation of sucrose into the medium [25]. However, green photoautotrophic callus cultures have been obtained from several different kinds of plants.

5. Study cases

Anthurium (*A. andreanum* Lind.) and moth orchids (*Phalaenopsis*is sp.) are tropical species with worldwide economic importance as ornamental plants and cut flowers. These species are commonly propagated by suckers; however, this propagation method is relatively slow and can cause disease transmission. Micropropagation has emerged as an alternative for fast mass production of *A. andreanum* and *Phalaenopsisis* plants of high phytosanitary quality.

For *A. andreanum*, nodal segments were excised from in vitro-derived adventitious shoots and were used as explants. For in vitro culture of *Phalaenopsis*is, protocorms were used as explants. The explants were placed in a 500 ml jar containing 40 ml of MS [26] medium without growth regulators. The pH of the culture medium was adjusted to 5.8 with 0.1 N sodium hydroxide, 0.25% (w/v) Phytagel was added as a gelling agent and then it was autoclaved for 15 min at 120°C and 117.7 kPa. The nodal segments were exposed to white LEDs (460 and 560 nm), red LEDs (660 nm), blue LEDs (460 nm), the combination of blue and red LEDs (460 and 660 nm, respectively), and FLs (545–610 nm) as a control. The LED system (model: 5050–1M-RGB, 3M, MN, USA) consisted of strips remotely controlled with a 12 V DC power adapter (model: SDK-0605, 3M, MN, USA). The explants were incubated at 24 ± 2°C and for 16 h light photoperiod. In all treatments, the photosynthetic photon flux density (PPFD) was maintained to 25 $\mu mol\ m^{-2}\ s^{-1}$. PPFD was measured using a FieldScout Quantum Light Meter®.

After 60 days of in vitro culture, shoot length (cm), number of leaves, rooted shoots, and chlorophyll *a*, chlorophyll *b*, and total chlorophyll contents were evaluated. Chlorophyll content was determined according to the method of Harborne [27]. For experimental design and data analysis, a completely randomized experimental design was used for all experiments. For each treatment, ten culture vessels, containing three explants each, were used. An analysis of variance (ANOVA) and Tukey's comparison of means test ($p \leq 0.05$) were performed for each species using SPSS statistical software (version 22 for Windows).

For *A. andreanum,* treatments with white LEDs, blue LEDs, and the combination of blue and red LEDs showed the greatest plantlet length and number of leaves. The FL and red LED treatments showed similar responses in promoting the formation of plantlets and their leaves. All shoots were rooted and the highest root number was induced in cultures incubated in FLs and blue LEDs with 6.6 and 6.0 roots, respectively. The lowest root number (1.5) was recorded in cultures incubated in red LEDs (**Table 1**). Chlorophyll a, b, and total chlorophyll content was significantly higher in the blue LED treatment (0.692 mg g^{-1} fresh weight), while the lowest total chlorophyll content was found in the red LED and FL treatments with 0.327 and 0.375 mg g^{-1} fresh weight, respectively (**Figure 3a**).

In *Phalaenopsisis,* treatments with FLs, white LED and the combination of blue and red LEDs showed the greatest plantlet length and number of leaves (**Table 1**) The white, red and blue LEDs showed similar responses in promoting the formation of plantlets and their leaves. All protocorms were rooted and had the same root number. Chlorophyll a content was significantly higher in the blue LED treatment (0.2813 mg g^{-1} fresh weight), while chlorophyll b content was higher in blue and the combination of blue and red LED treatments, with 0.1368 and 0.1468 mg g^{-1} fresh weight, respectively. Total chlorophyll (0.421875 mg g^{-1} fresh weight)

Treatment	Shoot length (cm)	No. of leaves	Rooting (%)	No. of roots
Anthurium andreanum				
Fluorescent lamps	3.1 ± 0.1 b	4.9 ± 0.5 b	100.0 ± 0.0 a	6.6 ± 0.3 a
White LEDs	4.3 ± 0.2 a	5.7 ± 0.4 ab	100.0 ± 0.0 a	4.3 ± 0.6 b
Red LEDs	2.9 ± 0.2 b	5.0 ± 0.3 b	100.0 ± 0.0 a	5.0 ± 0.2 c
Blue LEDs	4.4 ± 0.4 a	6.8 ± 0.5 a	100.0 ± 0.0 a	6.0 ± 0.4 ab
Blue + red LEDs	4.0 ± 0.3 a	5.5 ± 0.3 ab	100.0 ± 0.0 a	2.5 ± 0.4 c
Phalaenopsis sp.				
Fluorescent lamps	17.0 ± 0.9 *a*	2.7 ± 0.2 *a*	100.0 ± 0.0 *a*	1.7 ± 0.2 *a*
White LEDs	14.5 ± 0.8 *ab*	2.3 ± 0.2 *ab*	100.0 ± 0.0 *a*	1.3 ± 0.2 *a*
Red LEDs	11.6 ± 0.7 *b*	1.7 ± 0.2 *b*	100.0 ± 0.0 *a*	1.5 ± 0.2 *a*
Blue LEDs	12.0 ± 0.4 *b*	1.7 ± 0.2 *b*	100.0 ± 0.0 *a*	1.3 ± 0.2 *a*
Blue + red LEDs	17.3 ± 0.6 *a*	2.7 ± 0.2 *a*	100.0 ± 0.0 *a*	1.7 ± 0.2 *a*

Table 1. Effect of LEDs on in vitro growth and rooting of *Anthurium andreanum* cv. Rosa and *Phalaenopsis* sp after 60 days of culture.

was higher in blue LED. The lowest total chlorophyll content was found in FL treatments and white LEDs with 0.1810 and 0.2500 mg g^{-1} fresh weight, respectively (**Figure 3b**).

Our results indicate that FLs can be replaced by LEDs. The same effect was observed by Kurilčik et al. [10] and Lin et al. [28] during in vitro development of *Chrysanthemum* plantlets and *Dendrobium officinale* protocorms, respectively. In *Phalaenopsis*, LEDs had no effect on the number of roots, while in *A. andreanum* the highest number of roots was obtained in FLs and

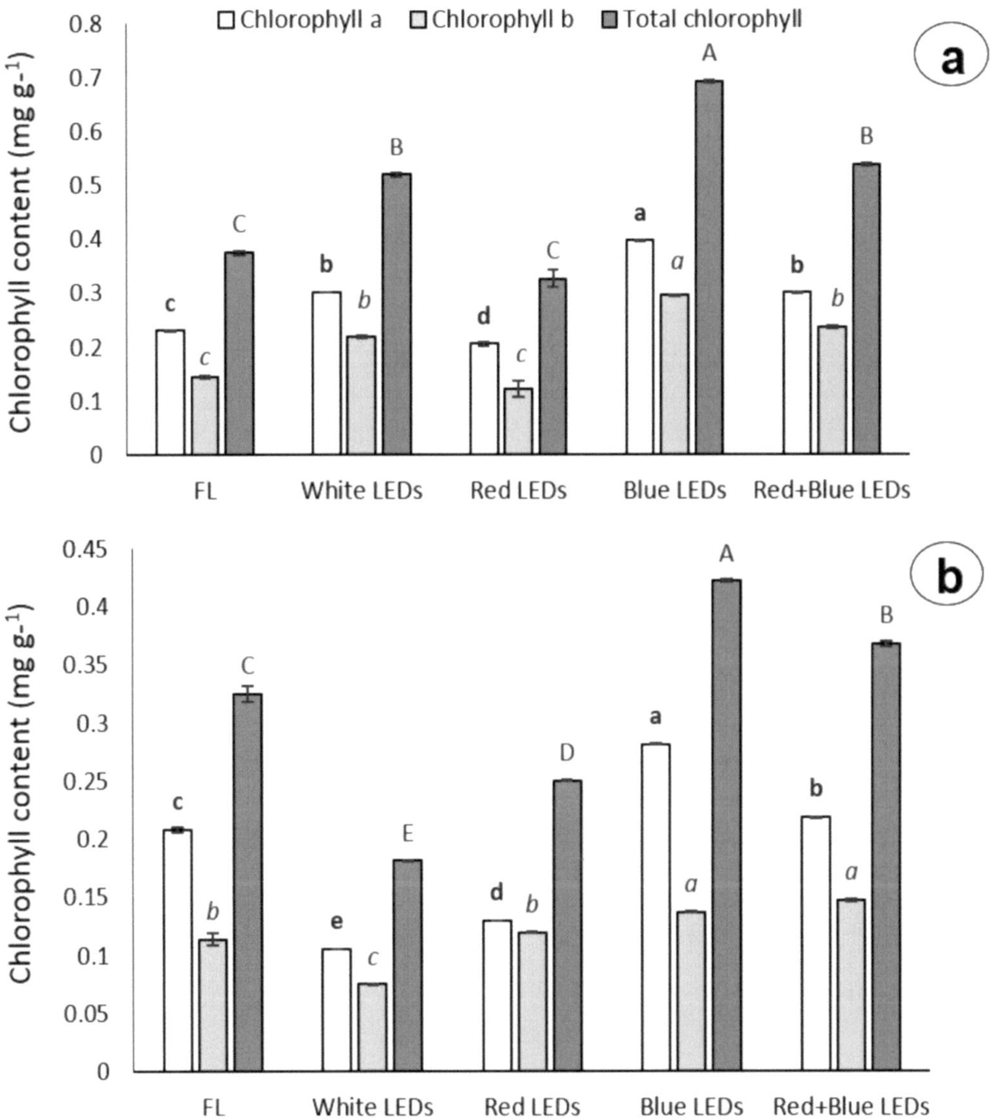

Figure 3. Effect of light quality on chlorophyll content in *Anthurium andreanum* (a) and *Phalaenopsis*is sp. (b) after 60 days of culture in vitro. Different letters denote statistically significant differences according to Tukey's multiple range test at $p \leq 0.05$. Bars represent mean ± SE.

blue LEDs. Similar results were reported by Cybularz-Urban et al. [29] and Waman et al. [7] in *Cattleya* and *Musa* spp., respectively.

According to Topchiy et al. [24], light quality also plays an important role in photosynthesis, influencing the way in which light is absorbed by chlorophyll. The present results demonstrated that the chlorophyll a, chlorophyll b, and total chlorophyll content appeared greater in plantlets growing under treatments containing blue light. Similar results were reported by Dewir et al. [15] where blue LEDs showed greater growth, vigor, and chlorophyll content in *E. milli*. Jao et al. [18] reported that blue LEDs promote growth and increase chlorophyll content in *Zantedeschia jucunda*. Our results are consistent with these studies in that the blue LEDs have an important role in the synthesis of photosynthetic pigments. This suggests that LEDs can also be used for improving the quality of *ex vitro* plantlets of *A. andreanum* and *Phalaenopsisis* sp.

In conclusion, the use of light-emitting diodes (LEDs) as a radiation source for plants has attracted considerable interest for commercial micropropagation. The flexibility of matching LED wavelengths to plant photoreceptors may provide more optimal production, influencing plant morphology, and chlorophyll content. Although previous reports have confirmed physiological and morphological effects of LED light quality on metabolism and development of several plantlets in vitro, in our experience, LED light is more suitable for plant morphogenesis and growth than FLs. However, the responses vary according to plant species.

Author details

Jericó J. Bello-Bello[1]*, Juan A. Pérez-Sato[1], Carlos A. Cruz-Cruz[2] and Eduardo Martínez-Estrada[1]

*Address all correspondence to: jericobello@gmail.com

1 Postgraduate College-Campus Córdoba, Amatlán de los Reyes, State of Veracruz, Mexico

2 Faculty of Chemical Sciences, University of Veracruz, Orizaba, State of Veracruz, Mexico

References

[1] Schuerger A. C., Brown C. S., Stryjewski E. C. Anatomical features of pepper plants (*Capsicum annuum* L.) grown under red light-emitting diodes supplemented with blue or far-red light. Annals of Botany. 1997;**79**(3):273–282.

[2] Tibbitts T. W., Morgan D. C., Warrington I. J. Growth of lettuce, spinach, mustard, and wheat plants under four combinations of high-pressure sodium, metal halide, and tungsten halogen lamps at equal PPFD. Journal of the American Society for Horticultural Science. 1983;**108**(4):622–630.

[3] Tennessen D. J., Singsaas E. L., Sharkey T. D. Light-emitting diodes as a light source for photosynthesis research. Photosynthesis Research. 1994;**39**(1):85–92.

[4] Tripathy B. C., Brown C. S. Root-shoot interaction in the greening of wheat seedlings grown under red light. Plant Physiology. 1995;**107**(2):407–411.

[5] Nhut D. T., Takamura T., Watanabe H., Okamoto K., Tanaka, M. Responses of strawberry plantlets cultured in vitro under superbright red and blue light-emitting diodes (LEDs). Plant Cell, Tissue and Organ Culture. 2003;**73**(1):43–52.

[6] Hung C. D., Hong C. H., Jung H. B., Kim S. K., Van Ket N., Nam M. W., et al. Growth and morphogenesis of encapsulated strawberry shoot tips under mixed LEDs. Scientia Horticulturae. 2015;**194**:194–200.

[7] Waman A. A., Bohra P., Sathyanarayana B. N., Umesha K., Gowda B., Ashok T. H. In vitro shoot multiplication and root induction in Silk banana variety Nanjanagud Rasabale as influenced by monochromatic light spectra. Proceedings of the National Academy of Sciences, India Section B: Biological Sciences. 2015;**86**(3):577–584.

[8] Jao R. C., Fang W. Effects of frequency and duty ratio on the growth of potato plantlets in vitro using light-emitting diodes. HortScience. 2004;**39**(2):375–379.

[9] Kim S. J., Hahn E. J., Heo J. W., Paek K. Y. Effects of LEDs on net photosynthetic rate, growth and leaf stomata of chrysanthemum plantlets in vitro. Scientia Horticulturae. 2004;**101**(1):143–151.

[10] Kurilčik A., Miklušyte-Čanova R., Dapkūniene S., Žilinskaite S., Kurilčik G., Tamulaitis G., et al. In vitro culture of chrysanthemum plantlets using light-emitting diodes. Central European Journal of Biology. 2008;**32**(2):161–167.

[11] Poudel P. R., Kataoka I., Mochioka R. Effect of red-and blue-light-emitting diodes on growth and morphogenesis of grapes. Plant Cell, Tissue and Organ Culture. 2008;**92**(2):147–153.

[12] Li H., Tang C., Xu Z. The effects of different light qualities on rapeseed (*Brassica napus* L.) plantlet growth and morphogenesis in vitro. Scientia Horticulturae. 2003;**150**:117–124.

[13] Kwon A. R., Cui H. Y., Lee H., Shin H., Kang K. S., Park S. Y. Light quality affects shoot regeneration, cell division, and wood formation in elite clones of *Populus euramericana*. Acta Physiologiae Plantarum. 2015;**37**(3):1–9.

[14] Ferreira L. T., de Araújo Silva M. M., Ulisses C., Camara T. R., Willadino L. Using LED lighting in somatic embryogenesis and micropropagation of an elite sugarcane variety and its effect on redox metabolism during acclimatization. Plant Cell, Tissue and Organ Culture. 2017;**128**(1):211–221.

[15] Dewir Y. H., Chakrabarty D., Kim S. J., Hahn E. J., Paek K. Y. Effect of light-emitting diode on growth and shoot proliferation of *Euphorbia millii* and *Spathiphyllum cannifolium*. Horticulture, Environment, and Biotechnology. 2005;**46**(6):375–379.

[16] Morrow R. C. LED lighting in horticulture. HortScience. 2008;**43**(7):1947–1950.

[17] Gupta S. D., Jatothu B. Fundamentals and applications of light-emitting diodes (LEDs) in in vitro plant growth and morphogenesis. Plant Biotechnology Reports. 2013;**7**(3):211–220.

[18] Jao R. C., Lai C. C., Fang W., Chang S. F. Effects of red light on the growth of Zantedeschia plantlets in vitro and tuber formation using light-emitting diodes. HortScience. 2005;**40**(2):436–438.

[19] Li H., Xu Z., Tang C. Effect of light-emitting diodes on growth and morphogenesis of upland cotton (*Gossypium hirsutum* L.) plantlets in vitro. Plant Cell, Tissue and Organ Culture. 2010;**103**(2):155–163.

[20] Li H., Tang C., Xu Z., Liu X., Han X. Effects of different light sources on the growth of non-heading Chinese cabbage (*Brassica campestris* L.). Journal of Agricultural Science. 2012;**4**(4):262–273.

[21] Moon H. K., Park S. Y., Kim Y. W., Kim C. S. Growth of Tsuru-rindo (*Tripterospermum japonicum*) culturedin vitro under various sources of light-emitting diode (LED) irradiation. Journal of Plant Biology. 2006;**49**(2):174–179.

[22] Soebo A., Krekling T., Appelgren M. Light quality affects photosynthesis and leaf anatomy of birch plantlets in vitro. Plant Cell, Tissue and Organ Culture. 1995;**41**(2):177–185.

[23] Goins G. D., Yorio N. C., Sanwo M. M., Brown C. S. Photomorphogenesis, photosynthesis, and seed yield of wheat plants grown under red light-emitting diodes (LEDs) with and without supplemental blue lighting. Journal of Experimental Botany. 1997;**48**(7):1407–1413.

[24] Topchiy N. M., Sytnik S. K., Syvash O. O., Zolotareva O. K. The effect of additional red irradiation on the photosynthetic apparatus of *Pisum sativum*. Photosynthetica. 2005;**43**(3):451–456.

[25] George E.F. The Background. In: George E.F., Hall M. A., De Klerk G. J., editors. Plant Tissue Culture Procedure – Background. Netherlands: Springer; 2008. pp. 1–28.

[26] Murashige T., Skoog F. A revised medium for rapid growth and bio assays with tobacco tissue. Physiologia Plantarum. 1962;**15**(3):473–497.

[27] Harborne J. B. Nitrogen compounds. In: Harbone A. J., editor. Phytochemical Methods. Netherlands: Springer; 1973. pp. 166–211.

[28] Lin Y., Li J., Li B., He T., Chun Z. Effects of light quality on growth and development of protocorm-like bodies of Dendrobium officinale in vitro. Plant Cell, Tissue and Organ Culture. 2011;**105**(3):329–335.

[29] Cybularz-Urban T., Hanus-Fajerska E., Swiderski A. Effect of light wavelength on in vitro organogenesis of a Cattleya hybrid. Acta Biologica Cracoviensia. 2007;**49**(1):113–118.

Chapter 7

Chlorophyll as Photosensitizer in Dye-Sensitized Solar Cells

Abdul Kariem Arof and Teo Li Ping

Additional information is available at the end of the chapter

http://dx.doi.org/10.5772/67955

Abstract

Chlorophyll, being the most abundant pigment that commonly found in plants, bacteria, bryophytes and algae, plays a vital role in photosynthesis. Chlorophylls are natural pigments and therefore safe, environmental friendly, easily available and cheap. Chlorophyll has been experimented to function as a photosensitizer in dye-sensitized solar cells (DSSCs) as DSSCs mimic the photosynthesis process in green plants. DSSC was first developed by Gratzel in 1991 and since then has gained tremendous attention as its fabrication is cheap and easy. A DSSC basically comprises a semiconductor that has been soaked in sensitizing dye (chlorophyll), a counter electrode, and an electrolyte containing a redox mediator. The dye absorbs light, which is transformed into electricity. Chlorophyll can be extracted from the leaves of pomegranate, bougainvillea, papaya, *Pandanus amaryllifolius*, spinach, green grasses, seaweeds, algae and bryophytes. Chlorophyll from these sources has been studied as possible photosensitizers for DSSCs. Most researches done in chlorophyll DSSC use the extracted natural pigments. The type of solvent and pH of the dye solution will also affect the stability of chlorophyll and subsequently the performance of the DSSCs. This chapter will present an inexhaustive overview on DSSCs using chlorophyll as dye.

Keywords: chlorophyll, photosensitizer, light adsorption, dye-sensitized solar cells, efficiency

1. Introduction

Over billions of years, Mother Nature has been converting light from the sun into energy via photosynthesis. Sunlight is the most abundant and sustainable energy source that is free.

The Earth receives energy from the sun at the rate of ~12×10^{17} J s^{-1} [1]. This has exceeded the yearly worldwide energy consumption rate of ~1.5×10^{13} J s^{-1} [1]. Therefore, it is a challenge to devise an approach for the effective capture and storage of solar energy for our consumption since fossil fuels such as oil and gas will be depleted in the years to come. In order to imitate the photosynthesis process, Gratzel and coworkers have developed dye-sensitized solar cells (DSSCs) based on the similar working mechanism [2]. Nevertheless, one main difference between photosynthesis of plants and DSSCs is that the energy can be stored in plants for later use but DSSC is unable to store energy. Ever since the birth of DSSCs, they have become the spotlight of attention among scientists and researchers around the world as they are much cheaper, easier to fabricate, and more environmental friendly when compared with conventional silicon solar cells [3, 4]. A DSSC is an electrochemical device that comprises a transparent-conducting oxide (TCO) glass over which is deposited a semiconductor. The semiconductor will be soaked in a dye solution. An electrolyte with reduction-oxidation (redox) mediator and cathode are the other remaining components. The fluorine-doped tin oxide (FTO)/semiconductor/dye assembly is referred to as photoanode. Indium-doped tin oxide (ITO) and FTO are two TCOs used commonly in DSSCs. Titanium dioxide (TiO_2) is one of the popular semiconductors used for DSSC since it is cheap, non-toxic, and possesses a large bandgap [5]. TiO_2 is deposited on the TCO substrate in the form of TiO_2 nanoporous particle network to increase the coverage area for the sensitizing dye. The cathode is made up of another TCO on top of which platinum is deposited. Carbon and conducting polymers can also be employed as counter electrode. If a gel polymer electrolyte is used, it is sandwiched between the photoanode and cathode. The dye, on the other hand, can be categorized into two groups: synthetic and natural. The most frequently used synthetic dye is the ruthenium (Ru)-based dyes but they are not environmental friendly since Ru is a heavy metal [6]. Such dyes are also very expensive due to the scarcity of Ru. By contrast, natural dyes are readily available and thus cheap besides being non-toxic, environmental friendly, biodegradable, easily extracted as well as can be used without any purification [6]. Since DSSC mimics the photosynthesis of green plants, therefore chlorophyll can also function as photosensitizer for DSSC. In fact, report on chlorophyll as photosensitizer on zinc oxide (ZnO) semiconductor was first published by Tributsch in 1972 [7].

2. Basic working principle of chlorophyll-sensitized DSSC

In this chapter, discussion is based on the TiO_2 semiconductor photoanode. However, occasionally we refer to zinc oxide (ZnO) and tin dioxide (SnO_2). The dye is chlorophyll extracted from various sources including leaves, grasses, flowers, seaweeds, and algae. The electrolyte is generally in the form of liquid and quasi-solid state. The commonly used mediator is the I^-/I_3^- redox couple and the counter electrode chosen in the preceding discussion is platinum (Pt) or carbon (C). Upon shining light on the cell, the molecules of the chlorophyll dye (D) will be excited (D^*) after absorbing photons (hv) and inject electrons into the semiconductor conduction band as described in the equation below:

http://dx.doi.org/10.5772/67955

$$hv + D \rightarrow D^* \tag{1}$$

The excited chlorophyll molecules (D^*) will inject electrons into the TiO_2 conduction band and the excited dye will then be oxidized or ionized (D^+). The reaction process involved is as follows:

$$D^* + TiO_2 \rightarrow D^+ + e^-_{cb}(TiO_2) \tag{2}$$

The oxidized chlorophyll dye molecules (D^+) will accept electrons from an iodide ion (I^-) in the electrolyte when the I^- ions were released to the oxidized molecules and in turn oxidized to triiodide ions (I_3^-) according to the equation below:

$$2D^+ + 3\,I^- \rightarrow I_3^- + 2D \tag{3}$$

The electron in the TiO_2 conduction band flows out of the device through the load to reach the counterelectrode and reduced the triiodide ion as follows:

$$I_3^- + 2\,e^- \rightarrow 3\,I^- \tag{4}$$

The iodide ion is now restored, the electron circuit is completed, and the whole system is back to its original state to start a new cycle. These processes will continue as long as there is light and current is produced in the external circuit continuously. Under illumination, the voltage generated is given by the energy difference between the photoanode's Fermi level and the electrolyte's redox potential. **Figure 1** illustrates the schematic diagram of the chlorophyll-sensitized DSSC and its operating principle.

The light to electricity conversion efficiency (η) of the chlorophyll DSSC can be calculated from the equation below:

$$\eta = \frac{J_{sc} \times V_{oc} \times FF}{P_{in}} \times 100\% \tag{5}$$

Here J_{sc} is the short circuit current density (unit: mA cm^{-2}), which is obtained without any external applied voltage or potential, V_{oc} is the open circuit voltage (unit: mV) obtained under the condition of open circuit when there is no current, P_{in} is the input power (total incident light power density), and FF is the fill factor which can be expressed as

$$FF = \frac{J_{max} \times V_{max}}{J_{sc} \times V_{oc}} \tag{6}$$

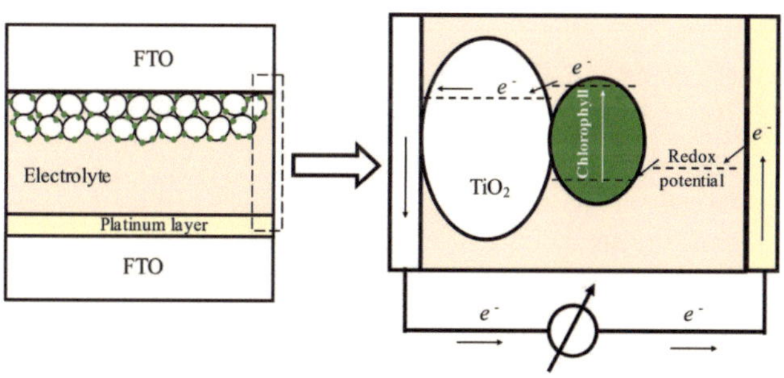

Figure 1. Operating principle of a chlorophyll-sensitized DSSC.

Here J_{max} and V_{max} are the photocurrent density and voltage in the *J-V* curve at the maximum power output. Each single component in a DSSC is important to ensure good performance. The main focus in this chapter is the chlorophyll dye. The dye is to absorb light, injects electrons into the semiconductor, and receives electrons from the redox mediator in the electrolyte. The cycle continues. An efficient dye sensitizer should display unique characteristics as listed below [8, 9]:

- Absorb light in the visible region.
- Good attachment at the surface of photoelectrode to ensure fast electron transfer.
- Good interfacial properties and high stability to enable good absorption to TiO_2.
- Easily accepting replacement electron from electrolyte.
- Excited state of dye must be slightly above the TiO_2 conduction band and its ground-state level below the redox potential of the electrolyte.
- Lifetime of the dye must be consistent with device life.
- Stable enough to sustain about 20 years exposure to natural light.

3. Performance of chlorophyll-sensitized DSSCs

Table 1 summarizes the performance of some DSSCs employing chlorophyll as photosensitizer reported by researchers worldwide. Herein, the illumination of the chlorophyll-sensitized DSSCs was carried out under intensity of 100 mW cm^{-2} unless stated otherwise.

It is evident that the condition of leaves whether fresh or dried affects the adsorption of chlorophyll onto the photoanode surface and consequently the performance. Taya et al. [10] observed that DSSCs having chlorophyll extracted from fresh leaves of *Anethum graveolens* (Indian traditional medicinal herb and spice) and arugula (arugula salad leaves) exhibited better performance than DSSCs with *A. graveolens* and arugula leaves that have been dried for 1 week. On the other hand, higher efficiencies were detected in DSSCs with dried parsley, spinach, and green algae as compared to fresh ones. However, the authors did not discuss the reason behind this. Among parsley, arugula, *A. graveolens*, *Spinach oleracea*, and green algae, chlorophyll extracted from spinach produced the best efficiency of 0.290 % [10]. The efficiency of the DSSCs depends on the soaking temperature and time. The 0.290% efficiency was obtained when the TiO_2 photoanode was soaked in the spinach extract solution at 60°C for 12 h [10]. Decreasing the temperature yielded low efficiency with η = ~0.0380% at 30°C. Beyond the optimum temperature, the efficiency decreased to ~0.175% (70°C) and ~0.0190% (80°C), respectively. Reducing the TiO_2 soaking time in the chlorophyll spinach solution to 2 h gave poor efficiency of ~0.021% [10]. Beyond 12 h of soaking, no obvious change in efficiency was observed from the spinach chlorophyll-sensitized DSSC [10]. Therefore, it can be inferred that the freshness of chlorophyll leaves, the soaking temperature of TiO_2 in chlorophyll solution, and its duration influenced the DSSC performance. From **Table 1**, comparison has been

Dye	Photoanode	Electrolyte	Counter electrode	J_{sc} (mA cm^{-2})	V_{oc} (V)	*FF*	η (%)	Ref.
Anethum graveolens leaves (fresh)	TiO_2/FTO	I^-/I_3^- LE	Pt/FTO	0.965	0.579	0.400	0.220	[10]
Anethum graveolens leaves (dried)	TiO_2/FTO	I^-/I_3^- LE	Pt/FTO	0.454	0.562	0.320	0.080	[10]
Arugula leaves (fresh)	TiO_2/FTO	I^-/I_3^- LE	Pt/FTO	0.788	0.599	0.420	0.200	[10]
Arugula leaves (dried)	TiO_2/FTO	I^-/I_3^- LE	Pt/FTO	0.713	0.594	0.430	0.180	[10]
Parsley leaves (fresh)	TiO_2/FTO	I^-/I_3^- LE	Pt/FTO	0.535	0.445	0.340	0.070	[10]
Parsley leaves (dried)	TiO_2/FTO	I^-/I_3^- LE	Pt/FTO	0.448	0.553	0.400	0.090	[10]
Bougainvillea spectabilis flower	TiO_2/ITO	I^-/I_3^- GPE	PEDOT/FTO	1.110	0.500	0.586	0.325	[11]
Amaranthus caudatus flower	TiO_2/ITO	I^-/I_3^- GPE	PEDOT/FTO	1.820	0.550	0.610	0.610	[11]
Cordyline fruticosa leaves	TiO_2	Not stated	Not stated	1.300	0.616	0.602	0.500	[12]
Pawpaw leaves	TiO_2/FTO	LE	Not stated	0.649	0.504	0.605	0.200	[13]
Pomegranate leaves	TiO_2/ITO	I^-/I_3^- LE	Pt/FTO	2.050	0.560	0.520	0.597	[14]
Platanus orientalis L. (Chinar leaves)	TiO_2/ITO	I^-/I_3^- LE	Pt/ITO	0.012	0.468	0.004	0.550	[15]
Shiso leaves	TiO_2/FTO	p-CuI	-	3.520	0.432	0.390	0.590	[16]
Bougainvillea leaves	Au/TiO_2/FTO	I^-/I_3^- LE	Pt/FTO	3.230	0.500	0.410	0.618	[17]
Ocimum Gratissimum (scent leaves)	TiO_2/FTO	I^-/I_3^- LE	Pt/FTO	0.044	0.466	0.400	0.021	[18]
Spinach oleracea	TiO_2/ITO	I^-/I_3^- LE	Pt/ITO	0.467	0.550	0.510	0.131	[19]
Spinach oleracea (fresh)	TiO_2/FTO	I^-/I_3^- LE	Pt/FTO	0.332	0.590	0.420	0.080	[10]
Spinach oleracea (dried)	TiO_2/FTO	I^-/I_3^- LE	Pt/FTO	1.110	0.583	0.460	0.290	[10]
	ZnO/FTO	I^-/I_3^- LE	Pt/FTO	0.123	0.226	0.200	0.008	[10]
Red spinach leaves	TiO_2/ITO	I^-/I_3^- LE	C/FTO	1.000	0.505	0.578	0.583	[20]
	TiO_2/ITO	I^-/I_3^- LE	C/FTO	0.700	0.559	0.455	0.357	[20]
	TiO_2/ITO	I^-/I_3^- LE	C/FTO	0.500	0.750	0.394	0.296	[20]
Green spinach leaves	ZnO/ITO	I^-/I_3^- LE	C/ITO	0.052	0.590	0.530	0.016	[21]

Dye	Photoanode	Electrolyte	Counter electrode	J_{sc} (mA cm^{-2})	V_{oc} (V)	*FF*	η (%)	Ref.
Papaya leaves	TiO_2/FTO	I^-/I_3^- LE	Pt/FTO	0.360	0.325	0.560	0.070	[22]
	TiO_2/ITO	I^-/I_3^- LE	C/FTO	0.060 mA	0.394	0.250	-	[23]
Jatropha leaves	TiO_2/ITO	I^-/I_3^- LE	C/FTO	0.042 mA	0.350	0.250	-	[23]
Ipomoea leaves extract	TiO_2/ITO	I^-/I_3^- LE	Pt/ITO	0.850	0.495	0.536	0.233	[19]
	TiO_2/ITO	I^-/I_3^- LE	Pt/ITO	0.914	0.540	0.563	0.278	[19]
	TiO_2/ITO	I^-/I_3^- LE	Pt/ITO	0.825	0.533	0.548	0.259	[19]
	TiO_2/ITO	I^-/I_3^- LE	Pt/ITO	1.120	0.565	0.592	0.318	[19]
	TiO_2/ITO	I^-/I_3^- LE	Pt/ITO	0.982	0.543	0.564	0.292	[19]
	TiO_2/ITO	I^-/I_3^- LE	Pt/ITO	0.915	0.510	0.552	0.253	[19]
Azadirachta indica (Neem) leaves	TiO_2/FTO	I^-/I_3^- LE	C/FTO	0.430	0.404	0.401	0.720	[24]
	TiO_2/FTO	I^-/I_3^- LE	Pt/FTO	0.230	0.467	0.392	0.050	[25]
Ziziphus jujuba leaves (dried)	TiO_2/FTO	I^-/I_3^- LE	Pt/FTO	3.180	0.652	0.519	1.077	[26]
Basil leaves (dried)	TiO_2/FTO	I^-/I_3^- LE	Pt/FTO	1.398	0.581	0.499	0.409	[26]
Basil flower	TiO_2/FTO	I^-/I_3^- LE	Pt/FTO	1.120	0.600	0.400	0.270	[27]
Mint flower	TiO_2/FTO	I^-/I_3^- LE	Pt/FTO	0.450	0.560	0.380	0.090	[27]
Mint leaves (dried)	TiO_2/FTO	I^-/I_3^- LE	Pt/FTO	0.980	0.579	0.400	0.227	[26]
Lemon leaves[a]	TiO_2/FTO	I^-/I_3^- LE	C/FTO	1.080	0.592	0.100	0.036	[28]
Morula leaves[a]	TiO_2/FTO	I^-/I_3^- LE	C/FTO	0.059	0.472	0.050	0.001	[28]
Fig leaves (dried)	TiO_2/FTO	I^-/I_3^- LE	Pt/FTO	2.091	0.596	0.515	0.642	[26]
Berry leaves (dried)	TiO_2/FTO	I^-/I_3^- LE	Pt/FTO	3.573	0.595	0.441	0.939	[26]
Pandanus amaryllifolius leaves	TiO_2/ITO	GPE	Pt/ITO	1.610	0.360	0.410	0.240	[29]
	TiO_2/FTO	GPE	Pt/FTO	1.190	0.490	0.630	0.390	[30]
	TiO_2/FTO	GPE	Pt/FTO	1.910	0.480	0.560	0.510	[31]

http://dx.doi.org/10.5772/67955

Dye	Photoanode	Electrolyte	Counter electrode	J_{sc} (mA cm^{-2})	V_{oc} (V)	*FF*	η (%)	Ref.
Banana leaves (dried)	TiO_2/FTO	I^-/I_3^- LE	Pt/FTO	1.770	0.596	0.492	0.522	[26]
Peach leaves (dried)	TiO_2/FTO	I^-/I_3^- LE	Pt/FTO	2.555	0.611	0.422	0.659	[26]
Black tea leaves	TiO_2/FTO	I^-/I_3^- LE	Pt/FTO	0.390	0.550	0.400	0.080	[27]
Coccinia indica leaves	SnO_2/FTO	I^-/I_3^- LE	Not stated	0.700	0.540	0.610	0.260	[32]
	La-SnO_2/FTO	I^-/I_3^- LE	Not stated	0.820	0.540	0.540	0.290	[32]
	La-Cu-SnO_2/FTO	I^-/I_3^- LE	Not stated	1.010	0.560	0.510	0.310	[32]
Ficus retusa Linn.	Au-TiO_2/ITO	$Ce^{4+/3+}$ LE	Pt/ITO	7.850	0.520	0.289	1.180	[33]
Garcinia suubelliptica	Au-TiO_2/ITO	$Ce^{4+/3+}$ LE	Pt/ITO	6.480	0.322	0.331	0.691	[33]
Perilla	TiO_2/FTO	I^-/I_3^- LE	Pt/FTO	1.360	0.522	0.696	0.500	[34]
Petunia	TiO_2/FTO	I^-/I_3^- LE	Pt/FTO	0.850	0.616	0.605	0.320	[34]
Eggplant pulp	TiO_2/FTO	I^-/I_3^- LE	Pt/FTO	0.350	0.630	0.390	0.090	[27]
Festuca ovina grass	TiO_2/ITO	I^-/I_3^- LE	C/ITO	1.189	0.548	0.699	0.460	[35]
Hierochloe odorata grass	TiO_2/FTO	I^-/I_3^- LE	Pt/FTO	2.199	0.594	0.355	0.460	[36]
Torulinium odoratum grass	TiO_2/FTO	I^-/I_3^- LE	Pt/FTO	1.004	0.654	0.483	0.320	[36]
Dactyloctenium aegyptium grass	TiO_2/FTO	I^-/I_3^- LE	Pt/FTO	0.698	0.719	0.481	0.240	[36]
Moss bryophyte *(hyophila involuta)*	TiO_2/FTO	GPE	Pt/FTO	5.780	0.600	0.570	1.970	[37]
	TiO_2/FTO	GPE	Pt/FTO	4.590	0.610	0.640	1.770	[38]
	TiO_2/FTO	GPE	Pt/FTO	5.960	0.580	0.580	2.000	[38]
	TiO_2/FTO	GPE	Pt/FTO	3.710	0.640	0.720	1.690	[38]
	TiO_2/FTO	GPE	Pt/FTO	5.370	0.550	0.730	2.170	[38]
	TiO_2/FTO	GPE	Pt/FTO	8.440	0.540	0.580	2.620	[38]
Rhoeo spathacea (Sw.) Stearn	Au-TiO_2/ITO	$Ce^{4+/3+}$ LE	Pt/ITO	10.900	0.496	0.274	1.490	[33]
Sargassum wightii (marine seaweed)[b]	ZnO/FTO	I^-/I_3^- LE	Pt/FTO	0.203	0.330	0.460	0.070	[39]

Dye	Photoanode	Electrolyte	Counter electrode	J_{sc} (mA cm^{-2})	V_{oc} (V)	*FF*	η (%)	Ref.
Kelp (brown algae)	TiO_2/TCO	I^-/I_3^- LE	Pt/TCO	0.433	0.441	0.620	-	[40]
Undaria pinnatifida (brown seaweed)	TiO_2/FTO	I^-/I_3^- LE	Pt/FTO	0.800	0.360	0.690	0.178	[41]
	TiO_2/FTO	I^-/I_3^- LE	Pt/FTO	10.700	0.530	0.600	3.400	[42]
	TiO_2/FTO	I^-/I_3^- LE	Pt/FTO	13.800	0.570	0.580	4.600	[42]
	TiO_2/FTO	I^-/I_3^- LE	Pt/FTO	8.600	0.470	0.600	2.500	[42]
	TiO_2/FTO	I^-/I_3^- LE	Pt/FTO	9.000	0.470	0.610	2.600	[42]
Cladophora sp. (green algae)	TiO_2/FTO	I^-/I_3^- LE	Pt/FTO	0.145	0.585	0.590	0.055	[43]
Green algae (fresh)	TiO_2/FTO	I^-/I_3^- LE	Pt/FTO	0.134	0.416	0.210	0.010	[10]
Green algae (dried)	TiO_2/FTO	I^-/I_3^- LE	Pt/FTO	0.397	0.559	0.440	0.100	[10]
Chlorella vulgaris (microalgae)	TiO_2/FTO	I^-/I_3^- LE	Pt/FTO	2.530	0.551	0.650	0.900	[44]

[a]Intensity 80 mW cm^{-2}.

[b]Intensity 45 mW cm^{-2}.

LE: liquid electrolyte; GPE: gel polymer electrolyte; PEDOT: poly(3,4-ethylenedioxythiophene).

Table 1. The photovoltaic performance of some chlorophyll-sensitized DSSCs.

made between TiO_2 and ZnO photoanodes for chlorophyll spinach DSSCs where the former gave better performance than the latter [10].

In the case of chlorophyll extract from ipomoea leaves (leaves of morning glory flower), 50°C is the optimum temperature for TiO_2 immersion with efficiency of 0.278% [19]. Lower efficiencies of 0.233 and 0.259% were obtained when the TiO_2-soaking temperature in ipomoea leaves extract solution were at 30 and 80°C, respectively for 24 h [19]. Other than temperature, pH of the dye solution is another factor influencing the efficiency. Maintaining the soaking temperature at 50°C, the pH of ipomoea dye solutions was adjusted to pH 1, 2, and 3 [19]. However, there was no mention on the type of acid used. Thus, it is not known whether the anion of acid had any influence on the DSSC performance. Nevertheless, improvement in efficiency can be seen when the acidity of the dye solution was adjusted to pH 1 and 2 with efficiencies of 0.318 and 0.292%, respectively. However, further increasing the pH to 3 decreased the efficiency ($\eta = 0.253\%$) [19].

The type of solvent used for pigments extraction can also give different results in the absorption spectrum [45–47]. From the work of Al-Alwani et al. [45], it has been reported that the UV-vis absorption spectra of chlorophyll extracted from *Pandanus amaryllifolius* (screwpine leaves) and *Cordyline fruticosa* (commonly known as Ti plant or cabbage palm) in ethanol and methanol solution displayed highest intensity absorption peaks among other solvents such as n-butyl alcohol, ethyl-acetate, n-hexane, chloroform, acetonitrile, ethyl-ether, and petroleum ether. Then, 1 g TiO_2 powder was added in chlorophyll extracted from *P. amaryllifolius* and *C. fruticosa* in respective solvents (ethanol for Pandanus leaves and methanol for *Cordyline* leaves) and water at different ratios. The addition of an appropriate quantity of water into the respective alcohol solvent increased the polarity of solution for better dye adsorption on the TiO_2 surface [45]. For solution containing TiO_2 and *P. amaryllifolius* chlorophyll, the best absorption spectrum was obtained at 2:1 of ethanol to water ratio, whereas the optimum ratio for TiO_2-*C. fruticosa* solution with mixture solvents of methanol and water was 3:1. Better dye absorption onto TiO_2 surface is said to improve the performance of DSSCs but there is no DSSC results in Ref. [45]. Nonetheless, in recent publication [12], the same group of authors have turned to response surface methodology (RSM) approach to investigate the various parameters involved in the chlorophyll extraction process from *C. fruticosa* simultaneously and then predict their response in order to obtain the optimized condition for its extraction. After taking three factors into consideration, that is, boiling temperature for organic solvents (acetonitrile, ethanol, and methanol), different pH ranging from 4 to 8 and temperature for chlorophyll extraction from 50 to 90°C, it was found that chlorophyll can be best extracted from the cordyline leaves under the condition where the solvent was ethanol (boiling point 78°C), pH of 7.99, and at temperature of 78.33°C [12]. As a result, the efficiency of 0.500% was obtained for DSSC with *C. fruticosa* chlorophyll as listed in **Table 1**.

It should be noted that both betalain and chlorophyll pigments can be extracted from the flowers of *Amaranthus caudatus* (common name: love-lies-bleeding, velvet flower) and *Bougainvillea spectabilis* using different solvents. Chlorophyll pigments have been obtained when 0.1 mol L^{-1} hydrochloric acid (HCl) was used to dissolve the amaranthus and bougainvillea flowers, whereas ethanol as solvent will yield betalain pigments from the same flowers. Surprisingly,

DSSCs having chlorophyll from amaranthus and bougainvillea flowers in **Table 1** demonstrated better performance as compared to DSSCs with betalain from the same flowers. The J_{sc}, V_{oc}, *FF*, and η values of betalain DSSC from *B. spectabilis* were 0.081 mA cm^{-2}, 0.450 V, 0.483, and 0.018%, respectively [11]. DSSCs using betalain extracted from *A. caudatus* flower gave the J_{sc} of 0.102 mA cm^{-2}, V_{oc} of 0.530 V, *FF* of 0.610, and η of 0.033% [11]. The efficiencies exhibited by chlorophyll-sensitized DSSCs from amaranthus (η = 0.610 %) and bougainvillea (η = 0. 325 %) flowers were surprisingly high considering that gel polymer electrolyte and PEDOT counter-electrode were used instead of conventional liquid electrolyte and Pt electrode [11].

Khan and coworkers [20] have examined the effect of acid treatment on TiO_2 nanoparticles in the making of TiO_2 paste to be coated on ITO glass substrate via the doctor blade method. It is found that chlorophyll from red spinach leaves-sensitized DSSC without any acid treatment on TiO_2 photoanode exhibited the efficiency of 0.296% which is lower than the TiO_2 acid-treated DSSC with chlorophyll extracted from the same source under intensity of 50 mW cm^{-2} [20]. The presence of acid can prevent agglomeration of TiO_2 nanoparticles and results in better TiO_2 dispersion and thereby offer more adsorption sites for the dye molecules [20]. Khan et al. [20] used citric acid (organic acid) and nitric acid (inorganic acid) to prevent TiO_2 agglomeration in the DSSC fabricated with chlorophyll from red spinach leaves, and TiO_2 treated with citric acid gave higher efficiency of 0.583% compared to that using nitric acid treatment on TiO_2 electrode (η = 0.357%). The lower efficiency yielded by nitric acid treatment could be due to nitric acid being a strong oxidizing acid and its corrosive nature may ruin the TiO_2 surface [48]. From **Table 1**, the J_{sc} and *FF* of the DSSC with chlorophyll extracted from red spinach leaves can also be seen to increase with efficiency following the order η (TiO_2) < η (nitric acid treated TiO_2) < η (citric acid treated TiO_2), whereas the V_{oc} values decreased in the same order.

It can be noted from **Table 1** that SnO_2 was employed as photoanode instead of TiO_2 in the cell having chlorophyll extracted from *Coccinia indica* (ivy gourd) leaves [32]. Comparison with TiO_2 revealed that SnO_2 is chemically stable and has larger bandgap of 3.6 eV and higher electron mobility [49, 50]. Due to its wide bandgap, it is less sensitive to UV degradation and thus possesses better stability as compared to TiO_2 [51]. Nonetheless, its large bandgap will also cause SnO_2 to have lower open circuit voltage. SnO_2 having high electron mobility can yield fast electron transport and therefore electron recombination can be decreased. From the table, the cells with chlorophyll from *C. indica* leaves exhibited the efficiencies of 0.260, 0.290, and 0.310% using three photoanodes, that is, SnO_2, La-doped SnO_2, and La-Cu-doped SnO_2, respectively [32]. Doping elements into DSSC photoanode improved the performance since more dye molecules can be adsorbed in the working electrode due to larger surface area owing to increased roughness and pores after doping [52].

Chang et al. [17] have investigated the plasmonic effect of gold (Au) nanoparticles with an average size of 27 nm in TiO_2 DSSC using chlorophyll from bougainvillea leaves. An efficiency of 0.618% was obtained. The Au nanoparticles showed localized surface plasmon resonance behavior when the frequency of the incident light came close to the surface plasmon frequency of Au and consequently improved light absorption leading to a considerably high efficiency of 0.618% as listed in **Table 1**. Also, the interface between Au and TiO_2 formed a Schottky barrier where electrons will be blocked from re-entering the dye or electrolyte, which decreased electron recombination and improved the DSSC performance. Earlier report on TiO_2 loaded with Au nanoparticles prior to chlorophyll sensitization was published by Lai and coworkers [33]. Instead of using I^-/I_3^- LE, the authors employed water-based electrolyte at ethanol:water ratio of 7:13 with $Ce^{4+}/^{3+}$ as redox couple since water-based DSSC can be totally free from toxic and is biologically friendly. The chlorophyll was extracted from herbal plant *Rhoeo spathacea* (Sw.) Stearn. Unexpectedly, an efficiency of more than 1% was produced by the chlorophyll *R. spathacea* water-based DSSC which is higher than that of a similar cell but with one of the earliest synthetic dyes, that is, crystal violet (η = 0.010%) [33]. The authors attributed the aggregation of crystal violet dyes as the culprit behind this based on its photocurrent density value (J_{sc} = 2.040 mA cm^{-2}), which is lower than that of *R. spathacea* cell (J_{sc} = 10.900 mA cm^{-2}) [33]. In fact, the other two water-based DSSCs using chlorophyll extracted from *Ficus retusa* Linn. (common name: bonsai plant) and *Garcinia subelliptica* (common name: happiness tree) have also demonstrated higher efficiency than that of crystal violet-sensitized DSSC with efficiencies of 1.180 and 0.691%, respectively [33].

From **Table 1**, it can be observed that most of the DSSCs employ liquid electrolytes based on I^-/I_3^- redox mediator. The maximum power conversion efficiencies of liquid electrolyte-based DSSCs using synthetic dyes have reached around 14% [53]. Liquid-based electrolytes are desired since they can infiltrate into the TiO_2 nanopores network to make contact with the dye molecules for dye regeneration. Still, liquid electrolyte-based cells have limited durability due to the possibility of leakage and volatility of solvents. Some of the solvents are flammable as well. In an attempt to avoid these complications, researchers worldwide are focusing on developing polymer-based electrolytes for DSSCs. Solid polymer electrolytes can exhibit reasonable ionic conductivities but have poor interfacial contact with electrodes. Hence, gel-type polymer electrolytes (GPEs) are being developed. GPEs, which are basically liquid electrolyte trapped in the polymer matrix, have good flexibility and conductivities comparable to those of liquid electrolytes. It can be clearly seen from **Table 1** that the DSSC using chlorophyll extracted from moss bryophyte and gel polymer electrolyte exhibited exceptionally high efficiencies of ~2–2.620% under different conditions. The bryophyte cell with η = 1.970% was obtained with GPE having polyacrylonitrile (PAN) as polymer host, tetrapropylammonium iodide (TPAI) salt, iodine, ethylene carbonate (EC), and propylene carbonate (PC) as solvent and plasticizer [37]. Efficiencies of 1.770 and 2.000% were attained for bryophyte cells using GPE based on poly(vinyl alcohol) (PVA) with single [potassium iodide (KI)] and double salts (KI and TPAI), respectively [38]. Higher efficiency and J_{sc} values observed in the cell having binary salts GPE could be most probably due to the higher number of iodide ions contained in the GPE. As for the second best performing chlorophyll bryophyte DSSC (η = 2.170 %), it is acquired using GPE based on poly(vinyl alcohol) (PVA) and double salts of potassium iodide (KI) and TPAI with the addition of 0.7 M tert-butylpyridine (TBP) [38]. TBP can be used either by incorporating it in electrolyte or photoanode in order to improve the V_{oc} and subsequently η. However, in this case where TBP was added into GPE, the TBP effect is insignificant as compared to the bryophyte cell when the working electrode was immersed in TBP for 1 h (η = 1.690%) [38]. Nonetheless, the efficiency of the latter is lower than the former owing to its lower J_{sc} value which might be due to lesser photon harvesting as a result of reflection and light scattering by TBP [38]. The most efficient chlorophyll bryophyte DSSC having η of 2.620% was attained when a co-adsorbent, that is, chenodeoxycholic acid (CDCA), was added in the moss bryophyte [38]. The GPE used was PVA-based double salt without TBP. CDCA served as spacer to prevent the self-aggregation of chlorophyll molecules, diminish electron recombination, and stabilize the chlorophyll, thereby improving the efficiency.

It is worth mentioning from **Table 1** that the cell having the efficiency of 0.590% with chlorophyll extracted from shiso leaves used copper iodide (CuI) as hole transport material (HTM) instead of conventional liquid electrolyte [16]. Therefore, the DSSC has the configuration of FTO/TiO_2/chlorophyll dye/CuI. CuI, a p-type semiconductor, has bandgap of 3.1 eV and good optical transparency [54, 55]. The p-CuI was coated onto the chlorophyll/TiO_2/FTO using dip- and spray-coating technique as this method involves low calcination temperature and thus the degradation of dye will not occur [16, 54]. The p-CuI solid-state DSSC has similar working principle with conventional DSSC except that after photon absorption, the dye molecules will be excited and then inject electrons and holes into TiO_2 and p-CuI, respectively. This indicates that the dye at ground state must be positioned below CuI valence band and the dye-excited state should be above the TiO_2 conduction band in order to ensure proper functioning of chlorophyll CuI DSSC. With the usage of HTM, there will be no issue on pigment deterioration since natural pigment is unstable against the oxidized species in electrolyte with iodine as redox mediator [16].

Most of the reports on chlorophyll-sensitized DSSCs summarized in **Table 1** do not contain information on the type of chlorophyll used. Among the six chlorophylls, chlorophyll *a*, which plays vital role in photosynthesis process, shows poor adsorption and sensitization on TiO_2 due to its structure that contains phytyl and alkyl groups causing steric hindrance that obstruct the chlorophyll molecules to bind efficiently with TiO_2 molecules [40, 56, 57]. The structure of chlorophyll *b* only differs from chlorophyll *a* by the aldehyde group (–CHO) rather than methyl group (–CH_3). On the other hand, chlorophyll *c*, which consists of chlorophyll c_1 and chlorophyll $c_{2,}$ has carboxyl group (–COOH) that can effectively attach to TiO_2 as reported in Ref. [41]. Chlorophyll *c*, which is the main pigment in *Undaria pinnatifida* (brown seaweed) yielded the efficiency of 0.178% when applied in DSSC as listed in **Table 1**. Wang et al. [42] have purified the pigments in *U. pinnatifida* to obtain chlorophyll *c* and remove chlorophyll *a* and carotenoids. Then, the purified chlorophyll *c* was subjected to polyethylene column chromatography to isolate chlorophyll c_1 and chlorophyll c_2. Using the same method, chlorophyll c_1 and chlorophyll c_2 in oxidized form can also be obtained. As a result, efficiencies of 3.400 and 4.600% have been obtained from the chlorophyll c_{1-} and chlorophyll c_2-sensitized DSSCs with liquid electrolyte [42]. Decrement in efficiency can be seen in DSSCs employing oxidized chlorophyll c_1 (designated as chlorophyll c'_1) and oxidized chlorophyll c_2 (chlorophyll c'_2) with the values of 2.500 and 2.600%, respectively [42]. Nonetheless, to the best of our knowledge, the chlorophyll-sensitized DSSC utilizing chlorophyll c_2 extracted from *U. pinnatifida* exhibited the highest efficiency among other chlorophyll-sensitized DSSCs till date. However, there remains the stability issue encountered by chlorophyll-sensitized DSSCs. Therefore, further work must be done to enhance the stability as well as improving the chlorophyll-sensitized DSSCs performance before they can be put into practical usage.

http://dx.doi.org/10.5772/67955

4. Summary

It has been shown that chlorophyll has good potential to serve as photosensitizer in dye-sensitized solar cells. Moreover, they are cheap, non-toxic, biodegradable, easily found, and easy to use as sensitizer. Although the efficiency is still considerably low with highest efficiency to date being only 4.600% from DSSC with *U. pinnatifida* chlorophyll c_2, there remains the possibility and room for improvement to further enhance the performance and improve stability of chlorophyll-sensitized DSSCs for practical applications.

Acknowledgements

The authors thank the University of Malaya and Malaysian Ministry of Higher Education (MOHE) for the UMRG grant no. RP024C-14AFR and PRGS grant no. PR001-2014A.

Author details

Abdul Kariem Arof* and Teo Li Ping

*Address all correspondence to: akarof@um.edu.my

Centre for Ionics University of Malaya, Physics Department, Faculty of Science, Kuala Lumpur, Malaysia

References

[1] Blankenship RE, Tiede DM, Barber J, Brudvig GW, Fleming G, Ghirardi M, Gunner MR, Junge W, Kramer DM, Melis A, Moore TA, Moser CC, Nocera DG, Nozik AJ, Ort DR, Parson WW, Prince RC, Sayre RT. Comparing photosynthetic and photovoltaic efficiencies and recognizing the potential for improvement. Science 2011;332:805–809.

[2] O' Regan B, Grätzel M. A low-cost, high efficiency solar cell based on dye-sensitized colloidal TiO_2 films. Nature 1991;353:737–740.

[3] Grätzel M. Photoelectrochemical cells. Nature 2001;414:338–344.

[4] Bella F, Bongiovanni R. Photoinduced polymerization: An innovative, powerful and environmentally friendly technique for the preparation of polymer electrolytes for dye sensitized solar. Review. Journal Photochemistry and Photobiology C 2013;16:1–21.

[5] Karruppuchamy S, Jeong, JM, Amalnerkar DP, Minoura H. Photoinduced hydrophilicity of titanium dioxide thin films prepared by cathodic electrodeposition. Vacuum 2006;80:494–498.

[6] Fernando JMRC and Senadeera GKR. Natural anthocyanins as photosensitizers for dye-sensitized solar devices. Current Science 2008;95:663–666.

[7] Tributsch H. Reaction of excited chlorophyll molecules at electrodes and in photosynthesis. Journal of Photochemistry and Photobiology 1972;16:261–269.

[8] Grätzel M. Dye-sensitized solar cells. Journal of Photochemistry and Photobiology C: Photochemistry Reviews 2003;4:145–153.

[9] Longo C and De Paoli M-A. Dye-sensitized solar cells: A successful combination of materials. Journal of Brazilian Chemistry Society 2003;14:889–901.

[10] Taya SA, El-Agez TM, El-Ghamri HS, Abdel-Latif MS. Dye-sensitized solar cells using fresh and dried natural dyes. International Journal of Materials Science and Applications. 2013;2:37–42.

[11] Godibo DJ, Anshebo ST, Anshebo TY. Dye sensitized solar cells using natural pigments from five plants and quasi-solid state electrolyte. Journal of Brazilian Chemical Society 2015;26:92–101.

[12] Al-Alwani MAM, Ludin NA, Mohamad AB, Kadhum AAH, Baabbad MM, Sopian K. Optimization of dye extraction from *Cordyline fruticosa* via response surface methodology to produce a natural sensitizer for dye-sensitized solar cells. Results in Physics 2016;6:520–529.

[13] Kimpa MI, Momoh M, Isah KU, Yahya HN, Ndamitso MM. Photoelectric characterization of dye sensitized solar cells using natural dye from pawpaw leaf and flame tree flower as sensitizers. Materials Science and Applications 2012;3:281–286.

[14] Chang H, Lo Y-J. Pomegranate leaves and mulberry fruit as natural sensitizers for dye-sensitized solar cells. Solar Energy 2010;84:1833–1837.

[15] Liu B-Q, Zhao X-P, Luo W. The synergistic effect of two photosynthetic pigments in dye-sensitized mesoporous TiO_2 solar cells. Dyes and Pigments 2008;76:327–331.

[16] Kumara GRA, Kaneko S, Okuya M, Onwona-Agyeman B, Konno A, Tennakone K. Shiso leaf pigments for dye-sensitized solid-state solar cell. Solar Energy and Solar Cells 2006;90:1220–1226.

[17] Chang H, Kao M-J, Chen, T-L, Kuo H-G. Natural sensitizer for dye-sensitized solar cells using a three layers of photoelectrode thin films with a Schottky barrier. American Journal of Engineering and Applied Science 2011;4:214–222.

[18] Eli D, Musa GP, Ezra D. Chlorophyll and betalain as light-harvesting pigments for nanostructured TiO_2 based dye-sensitized solar cells. Journal of Energy and Natural Resources. 2016;5:53–58.

[19] Chang H, Wu HM, Chen TL, Huang KD, Jwo CS, Lo YJ. Dye-sensitized solar cell using natural dyes extracted from spinach and ipomoea. Journal of Alloys and Compounds 2010;495:606–610.

[20] Khan MA, Khan SMM, Mohammed MA, Sultana S, Islam JMM, Uddin J. Sensitization of nanocrystalline titanium dioxide solar cells using natural dyes: Influence of acids medium on coating formulation. American Academic & Scholarly Research Journal. 2012;4:10.

[21] Hasoon SA, Al-Haddad RMS, Shakir OT, Ibrahim IM. Natural dye-sensitized solar cell based on zinc oxide. International Journal of Scientific & Engineering Research. 2015;6:137–142.

[22] Suyitno S, Saputra TJ, Supriyanto A, Arifin Z. Stability and efficiency of dye-sensitized solar cells based on papaya-leaf dye. Spectrochimica Acta Part A: Molecular and Biomolecular Spectroscopy 2015;148:99–104.

[23] Pramono SH, Maulana E, Prayogo AF, Djatmika R. Characterization of dye-sensitized solar cell (DSSC) based on chlorophyll dye. International Journal of Applied Engineering Research 2015;10:193–205.

[24] Swarnkar AK, Sahare S, Chander N, Gangwar RK, Bhoraskar SV, Bhave TM. Nanocrystalline titanium dioxide sensitised with natural dyes for eco-friendly solar cell application. Journal of Experimental Nanoscience 2015;10:1001–1011.

[25] Rosana NTM, Amarnath DJ, Anandan S, Saritha G. Environmental friendly photosensitizing materials for harvesting solar energy. Journal of Materials and Environmental Science 2015;6:2053–2059.

[26] Taya SA, El-Agez TM, Elrefi KS, Abdel-Latif MS. Dye-sensitized solar cells based on dyes extracted from dried plant leaves. Turkish Journal of Physics 2015;39:24–30

[27] Abdel-Latif MS, Abuiriban MB, Dahoudi NA, Al-Kahlout AM, Taya SA, El-Agez TM, El-Ghamri HS. Dye-sensitized solar cells using fifteen natural dyes as sensitizers of nanocrystalline TiO_2. Science, Technology and Development 2015;34:135–139.

[28] Maabong K, Muiva CM, Monowe P, Sathiaraj TS, Hopkins M, Nguyen L, Malungwa K, Thobega M. Natural pigments as photosensitizers for dye-sensitized solar cells with TiO_2 thin films. International Journal of Renewable Energy Research. 2015;5:54–60.

[29] Noor MM, Buraidah MH, Yusuf SNF, Careem MA, Majid SR, Arof AK. Performance of dye-sensitized solar cells with (PVDF-HFP)-KI-EC-PC electrolyte and different dye materials. International Journal of Photoenergy 2011;2011:5 pages, Article ID 960487.

[30] Shah S, Buraidah MH, Teo LP, Careem MA, Arof AK. Dye-sensitized solar cells with sequentially deposited anthocyanin and chlorophyll as sensitizers. Optical and Quantum Electronics 2016;46:219, 8 pages.

[31] Noor MM, Buraidah MH, Careem MA, Majid SR, Arof AK. An optimized poly(vinylidene fluoride-hexafluoropropylene)-NaI gel polymer electrolyte and its application in natural dye sensitized solar cells. Electrochimica Acta 2014;121:159–167.

[32] Priyadharsini CI, Prakasam A, Anbarasan PM. Enhanced photoelectrochemical performance of DSSC by co-doped SnO_2 nanoparticles. International Letters of Chemistry, Physics and Astronomy 2013;17:82–93.

[33] Lai WH, Su YH, Teoh LG, Hon MH. Commercial and natural dyes as photosensitizers for a water-based dye-sensitized solar cell loaded with gold nanoparticles. Journal of Photochemistry and Photobiology A: Chemistry 2008;195:307–313.

[34] Zhou H, Wu L, Gao Y, Ma T. Dye-sensitized solar cells using 20 natural dyes as sensitizers. Journal of Photochemistry and Photobiology A: Chemistry 2011;219:188–194.

[35] Hernandez-Martinez AR, Estevez M, Vargas S, Quintanilla F, Rodriguez R. Natural pigments-based dye-sensitized solar cells. Journal of Applied Research and Technology 2012;10:38–47.

[36] Shanmugan V, Manoharan S, Sharafali A, Anandan S, Murugan R. Green grasses as light harvesters in dye sensitized solar cells. Spectrochimica Acta Part A: Molecular and Biomolecular Spectroscopy 2015;135:947–952.

[37] Hassan HC, Abidin ZHZ, Careem MA, Arof AK. Chlorophyll as sensitizer in I^-/I_3^--based solar cells with quasi-solid-state electrolytes. High Performance Polymers 2014;26:647–652.

[38] Hassan HC, Abidin ZHZ, Chowdhury FI, Arof AK. A high efficiency chlorophyll sensitized solar cell with quasi solid PVA based electrolyte. International Journal of Photoenergy 2016;2016:9 pages, Article ID 3685210.

[39] Anand M, Suresh S. Marine seaweed Sargassum wightii extract as a low-cost sensitizer for ZnO photoanode based dye-sensitized solar cell. Advances in Natural Sciences: Nanoscience and Nanotechnology 2015;6:035008, 8 pages.

[40] Hao S, Wu J, Huang Y, Lin J. Natural dyes as photosensitizers for dye-sensitized solar cells. Solar Energy 2006;80:209-214.

[41] Calogero G, Citro I, Di Marco G, Minicante SA, Morabito M, Genovese G. Brown seaweed pigment as a dye source for photoelectrochemical solar cells. Spectrochimica Acta Part A: Molecular and Biomolecular Spectroscopy 2014;117:702–706.

[42] Wang X-F, Zhan C-H, Maoka T, Wada Y, Koyama Y. Fabrication of dye-sensitized solar cells using chlorophylls c_1 and c_2 and their oxidized forms c'_1 and c'_2 from *Undaria pinnatifida* (Wakame). Chemical Physics Letters 2007;447:79–85.

[43] Lim A, Manaf NH, Tennakoon K, Chandrakanthi RLN, Lim LBL, Bandara JMRS, Ekanayake P. Higher performance of DSSC with dyes from *Cladophora* sp. As mixed cosensitizer through synergistic effect. Journal of Biophysics 2015;2015:8 pages, Article ID 510467.

[44] Mohammadpour R, Janfaza S, Abbaspour-Aghdam F. Light harvesting and photocurrent generation by nanostructured photoelectrodes sensitized with a photosynthetic pigment: A new application for microalgae. Bioresource Technology 2014;163:1–5.

[45] Al-Alwani MAM, Mohamad AB, Kadhum AAH, Ludin NA. Effect of solvents on the extraction of natural pigments and adsorption onto TiO_2 for dye-sensitized solar cell applications. Spectrochimica Acta Part A: Molecular and Biomolecular Spectroscopy 2015;138:130–137.

[46] Ayalew WA, Ayele DW. Dye-sensitized solar cells using natural dye as light-harvesting materials extracted from *Acanthus sennii chiovenda* flower and *Euphorbia cotinifolia* leaf. Journal of Science: Advanced Materials and Devices 2016;1:488–494.

[47] Mohammed IK, Kasim UI, Yabagi JA, Taufiq S. The effect on extracting solvents using natural dye extracts from *Hyphaene thebaica* for dye-sensitized solar cells. Journal of Material Science & Engineering 2015;4:208, 3 pages.

[48] Hao S, Wu J, Fan L, Huang Y, Lin J, Wei Y. The influence of acid treatment of TiO_2 porous film electrode on photoelectric performance of dye-sensitized solar cell. Solar Energy 2004;76:745–750.

[49] Dou X, Sabba D, Mathews N, Wong LH, Lam YM, Mhaisalkar S. Hydrothermal synthesis of high electron mobility Zn-doped SnO_2 nanoflowers as photoanode material for efficient dye-sensitized solar cells. Chemistry of Materials 2011;23:3938–3945.

[50] Mao X, Zhou R, Zhang S, Ding L, Wan L, Qin S, Chen Z, Xu J, Miao S. High efficiency dye-sensitized solar cells constructed with composites of TiO_2 and the hot-bubbling synthesized ultra-small SnO_2 nanocrystals. Scientific Reports 2016;6:19390. doi: 10.1038/srep19390 (2016)

[51] Bhande SS, Shinde DV, Tehare KK, Patil SA, Mane RS, Naushad M, Alothman ZA, Hui KN, Han SH. DSSCs synergic effect in thin metal oxide layer-functionalized SnO_2 photoanodes. Journal of Photochemistry and Photobiology A: Chemistry 2014;295:64–69.

[52] Arunachalam A, Dhanapandian S, Manoharan C. Effect of Sn doping on the structural, optical and electrical properties of TiO_2 films prepared by spray pyrolysis. Physica E 2016;76:35–46.

[53] Kakiage K, Aoyama Y, Yano T, Oya K, Fujisawa J-i, Hanaya M. Highly-efficient dye-sensitized solar cells with collaborative sensitization by silyl-anchor and carboxy-anchor dyes. Chemical Communications 2015;51:15894–15897.

[54] Tennakone K, Kumara GRRA, Kumarasinghe ARK, Wijayantha KGU, Sirimanne PM. Dye-sensitized nano-porous solid-state photovoltaic cell. Semiconductor Science and Technology 1995;10:1689–1693.

[55] Amalina MN, Rasheid NA, Rusop M. The properties of sprayed nanostructured p-type CuI films for dye-sensitized solar cells application. Journal of Nanomaterials 2012:2012:6 pages, Article ID 637637.

[56] Kay A, Grätzel M. Artificial photosynthesis I. Photosensitization of TiO_2 solar cells with chlorophyll derivatives and related natural porphyrins. The Journal of Physical Chemistry 1993;97:6272–6277.

[57] Calogero G, Bartolotta A, Di Marco G, Di Carlo A, Bonaccorso F. Vegetable-based dye-sensitized solar cells. Chemical Society Reviews 2015;44:3244–3294.